Springer Tracts in Modern Physics
Volume 235

Springer Tracts in Modern Physics

Springer Tracts in Modern Physics provides comprehensive and critical reviews of topics of current interest in physics. The following fields are emphasized: elementary particle physics, solid-state physics, complex systems, and fundamental astrophysics.
Suitable reviews of other fields can also be accepted. The editors encourage prospective authors to correspond with them in advance of submitting an article. For reviews of topics belonging to the above mentioned fields, they should address the responsible editor, otherwise the managing editor.
See also springer.com

Arno Straessner

Electroweak Physics at LEP and LHC

Jun. Prof. Dr. Arno Straessner
TU Dresden
Inst. für Kern- und Teilchenphysik
(IKTP)
Zellescher Weg 19
01069 Dresden
Germany
Arno.Straessner@cern.ch

A. Straessner, *Electroweak Physics at LEP and LHC*, STMP 235 (Springer, Berlin Heidelberg 2010), DOI 10.1007/978-3-642-05169-2

ISSN 0081-3869 e-ISSN 1615-0430
ISBN 978-3-642-05168-5 e-ISBN 978-3-642-05169-2
DOI 10.1007/978-3-642-05169-2
Springer Heidelberg Dordrecht London New York

Library of Congress Control Number: 2009943514

Cover design: Integra Software Services Pvt. Ltd., Pondicherry

Printed on acid-free paper

Springer is part of Springer Science+Business Media (www.springer.com)

To my wife Marlis

"Toujours ta montagne, grand fou!
Pense un peu à autre chose."

Roger Frison-Roche, Premier de Cordée

Preface

During more than 10 years, from 1989 until 2000, the LEP accelerator and the four LEP experiments, ALEPH, DELPHI, L3 and OPAL, have taken data for a large amount of measurements at the frontier of particle physics. The main outcome is a thorough and successful test of the Standard Model of electroweak interactions. Mass and width of the Z and W bosons were measured precisely, as well as the Z and photon couplings to fermions and the couplings among gauge bosons.

The first part of this work will describe the most important physics results of the LEP experiments. Emphasis is put on the properties of the W boson, which was my main research field at LEP. Especially the precise determination of its mass and its couplings to the other gauge bosons will be described. Details on physics effects like Colour Reconnection and Bose-Einstein Correlations in W-pair events shall be discussed as well. A conclusive summary of the current electroweak measurements, including low-energy results, as the pillars of possible future findings will be given. The important contributions from Tevatron, like the measurement of the top quark and W mass, will round up the present day picture of electroweak particle physics.

In the Standard Model, the close relationship between W and Z masses and the electroweak couplings is a consequence of the Higgs mechanism and electroweak symmetry breaking. This mechanism provides gauge invariant mass terms for all known elementary particles. The spectrum of particles is however extended by a scalar Higgs boson which has not been observed, yet. At LEP and at the Tevatron collider, searches for this particle were up to now not successful. The hunt for the Standard Model Higgs boson is therefore one of the main activities at future experiments. A new era will begin with the operation of the LHC collider. The ATLAS and CMS experiments have the potential to discover the Higgs boson in all theoretically possible mass ranges.

The second part of this volume will introduce the expected electroweak measurements as well as Higgs searches at the LHC. The experimental tools of the ATLAS and CMS detectors for the various measurements are described. At the LHC, the mass of the W boson and of the top quark will be determined with even greater precision than today's measurements. There is also the opportunity to improve the knowledge about the weak mixing angle and the triple gauge boson couplings.

One of the primary goals of the LHC experiments is the search for the Standard Model Higgs boson. The identification of the Higgs is summarised together with the measurement of its fundamental properties like its spin and behaviour under CP transformation, which will possibly be subject of future research. Eventually, conclusions and an outlook to possible future findings at the LHC will be given.

The measurements and the knowledge about particle physics presented in this work reflect the status of Summer 2009. It is expected that there will be new, maybe surprising findings in the near future. The electroweak data will however remain the cornerstone of particle physics to which new theories always need to be compared to.

Geneva and Dresden, September 2009 *Arno Straessner*

Contents

Chapter 1
Theoretical Framework

The currently known spectrum of elementary particles consists of leptons and quarks, which constitute the different forms of matter, and vector bosons, which are the force carriers.[1] The leptons appear in three families (ν_e, e), (ν_μ, μ), (ν_τ, τ), as well as the quarks (u, d), (c, s), (t, b). In the Standard Model [1, 2], forces between these elementary fermions are due to a $SU(3) \times SU(2) \times U(1)$ gauge symmetry of the corresponding field theory. The $SU(2) \times U(1)$ symmetry is generating the electroweak forces, with the photon, W and Z gauge bosons. The strong force is due to the $SU(3)$ symmetry. Quantum Chromodynamics [2] (QCD) describe the interaction of quarks and the corresponding gauge bosons, the gluons. In the following, natural units, setting $c = \hbar = 1$, are chosen, and the relations $c\hbar = 197.3269631(49)$ MeV fm [3] and $c = 299{,}792{,}458\,\mathrm{ms}^{-1}$ may be used to convert between energy and space-time units.

1.1 Electroweak Interactions

The electroweak part of the Standard Model Lagrangian can be divided into three parts, a gauge boson, a fermion and a Higgs term:

$$\mathcal{L} = \mathcal{L}_G + \mathcal{L}_F + \mathcal{L}_H \tag{1.1}$$

The $SU(2)$ and $U(1)$ gauge boson fields are $\mathbf{W}_\mu$ and B_μ. They couple to the weak isospin T_a and the weak hypercharge Y of the fermions. Left-handed fermion fields $\psi_L = \frac{1}{2}(1 - \gamma_5)\psi$ are combined to iso-doublets. The right-handed fields $\psi_R = \frac{1}{2}(1 + \gamma_5)\psi$ are iso-singlets. The corresponding values of the third component of the isospin, T_3, and Y are listed in Table 1.1, together with the electric charge Q. The left-handed down-type quarks, (d', s', b'), are related to their mass eigenstates, (d,s,b), by the Cabibbo-Kobayashi-Maskawa (CKM) mixing matrix [4, 3] according to $d'_i = \sum_{ij} V_{ij}^{\mathrm{CKM}} d_j$.

[1] Effects of gravity are too small to be observed in the energy ranges discussed here, and are neglected.

A. Straessner, *Electroweak Physics at LEP and LHC*, STMP 235, 1–43,
DOI 10.1007/978-3-642-05169-2_1, © Springer-Verlag Berlin Heidelberg 2010

Table 1.1 Quantum numbers of leptons and quarks. They are the eigenvalues of the third component of the weak isospin, T_3, of the weak hyper-charge, Y, and of the electrical charge, Q. The doublets of the weak isospin are put in *brackets*

Fermion Type			T_3	Y	Q
$\begin{pmatrix} \nu_e \\ e \end{pmatrix}_L$	$\begin{pmatrix} \nu_\mu \\ \mu \end{pmatrix}_L$	$\begin{pmatrix} \nu_\tau \\ \tau \end{pmatrix}_L$	$1/2$ $-1/2$	$-1/2$ $-1/2$	0 -1
$\nu_{e,R}$	$\nu_{\mu,R}$	$\nu_{\tau,R}$	0	0	0
e_R	μ_R	τ_R	0	-1	-1
$\begin{pmatrix} u \\ d' \end{pmatrix}_L$	$\begin{pmatrix} c \\ s' \end{pmatrix}_L$	$\begin{pmatrix} t \\ b' \end{pmatrix}_L$	$1/2$ $-1/2$	$1/6$ $1/6$	$2/3$ $-1/3$
u_R	c_R	t_R	0	$2/3$	$2/3$
d_R	s_R	b_R	0	$-1/3$	$-1/3$

The gauge part of the Lagrangian is given by:

$$\mathcal{L}_G = -\frac{1}{4} F_i^{\mu\nu} F_{\mu\nu}^i - \frac{1}{4} B_{\mu\nu} B^{\mu\nu} , \tag{1.2}$$

where $F_{\mu\nu}^i$ is the $SU(2)$ field strength

$$F_{\mu\nu}^i = \partial_\mu W_\nu^i - \partial_\nu W_\mu^i - g_2 \varepsilon_{ijk} W_\mu^j W_\nu^k \tag{1.3}$$

with the coupling constant g_2, and $B_{\mu\nu}$ the $U(1)$ field strength

$$B_{\mu\nu} = \partial_\mu B_\nu - \partial_\nu B_\mu . \tag{1.4}$$

The totally anti-symmetric tensor ε_{ijk} is identical to the $SU(2)$ structure constants. Due to the non-abelian $SU(2)$ group structure the W^i gauge fields do not evolve independently but are coupled to each other.

The interaction between fermions and gauge bosons is most conveniently written by means of the covariant derivate

$$D_\mu = \partial_\mu + i\frac{g_1}{2} Y B_\mu + i g_2 T_a W_\mu^a \tag{1.5}$$

yielding

$$\mathcal{L}_F = \sum_f i \bar{\psi}_f D_\mu \gamma^\mu \psi_f , \tag{1.6}$$

where the sum extends over all fermion fields. The T_a matrices are the two-dimensional representation of the group generators of the $SU(2)$, which follow the commutation relations $[T_i, T_j] = i\varepsilon_{ijk} T_k$ and $[T_i, Y] = 0$.

In the Standard Model, gauge invariant mass terms for fermions and bosons arise through the coupling to a complex doublet $\phi = \begin{pmatrix} \phi_1 \\ \phi_2 \end{pmatrix}$ of spin-zero Higgs fields and the spontaneous breaking of the $SU(2) \times U(1)$ symmetry [5]. In the minimal version there is only one Higgs doublet. The $\mathcal{L}_H$ term is completed by a dynamic term, a Higgs potential, and mass terms for the fermion fields:

$$\mathcal{L}_H = D_\mu \phi^\dagger D^\mu \phi - V(\phi) + \sum_f c_f \left(\bar{\psi}_f^L \phi^\dagger \psi_f^R + \bar{\psi}_f^R \phi \psi_f^L \right) \tag{1.7}$$

The ground state $\langle \phi \rangle_0$ of the Higgs self-interaction potential

$$V(\phi) = \mu^2 \phi^\dagger \phi + \lambda (\phi^\dagger \phi)^2 \tag{1.8}$$

is found for

$$\langle \phi^\dagger \phi \rangle_0 = \frac{v^2}{2} \tag{1.9}$$

with

$$v \equiv \sqrt{\frac{-\mu^2}{\lambda}} \,. \tag{1.10}$$

The Higgs field is rotated so that only the lower component remains and is then developed around the vacuum expectation value:

$$\langle \phi \rangle_0 = \frac{1}{\sqrt{2}} \begin{pmatrix} 0 \\ v + H \end{pmatrix} \,. \tag{1.11}$$

This choice breaks the original $SU(2) \times U(1)$ symmetry but conserves the electric charge symmetry, $U(1)_{\text{QED}}$. The energy scale v is not predicted by the model and must be measured experimentally.

With the charged vector boson fields

$$W_\mu^\pm = \frac{1}{\sqrt{2}} \left(W_\mu^1 \mp i W_\mu^2 \right) \tag{1.12}$$

the particle mass terms are given by

$$\begin{aligned} \mathcal{L}_{\text{mass}} = & -\frac{v}{\sqrt{2}} \sum_f c_f \bar{\psi}_f \psi_f \\ & + \left(\frac{v g_2}{2} \right)^2 W_\mu^+ W_-^\mu + \frac{v^2}{8} \left(W_\mu^3, B_\mu \right) \begin{pmatrix} g_2^2 & -g_1 g_2 \\ -g_1 g_2 & g_1^2 \end{pmatrix} \begin{pmatrix} W^{3,\mu} \\ B^\mu \end{pmatrix} \\ & + v^2 \lambda H^2 \end{aligned} \tag{1.13}$$

The Higgs boson mass depends on both v and the free parameter λ:

$$M_{\mathrm{H}} = v\sqrt{2\lambda}\,. \tag{1.14}$$

The fermion masses turn out to be

$$m_{\mathrm{f}} = \frac{v}{\sqrt{2}} c_f \tag{1.15}$$

with free Yukawa coupling constants c_f, not constrained by the model.

The measurements of neutrino flavour oscillations [6] clearly show that neutrinos are not massless, opposed to the original version of the Standard Model [1]. In the given formalism, neutrinos can be treated in the same way as the charged leptons. Like for quarks, the neutrinos are in general not identical to the mass eigenstates and the Pontecorvo-Maki-Nakgawa-Sakata (PMNS) mixing matrix needs to be added [7]. The resulting Dirac mass term is however not the only possibility to achieve massive neutrinos. An alternative is a Majorana mass term:

$$\mathcal{L}_{\mathrm{Majorana}} = -m_M \left(\bar{\psi}^c_{\nu,L} \psi_{\nu,L} + \mathrm{h.c.}\right) \tag{1.16}$$

However, this contribution does not conserve the lepton number and may give rise to a neutrino-less nuclear double beta decay, which is not observed, yet [8].

The symmetry breaking induces a mixing of the neutral boson fields W^3 and B, which can be diagonalised by

$$Z_\mu = \cos\theta_{\mathrm{w}} W^3_\mu - \sin\theta_{\mathrm{w}} B_\mu \tag{1.17}$$

$$A_\mu = \sin\theta_{\mathrm{w}} W^3_\mu + \cos\theta_{\mathrm{w}} B_\mu\,, \tag{1.18}$$

where the physical photon and Z boson fields, A_μ and Z_μ, appear. The weak mixing angle θ_{w} is defined by the ratio of the coupling constants g_1 and g_2:

$$\tan\theta_{\mathrm{w}} = \frac{g_1}{g_2} \tag{1.19}$$

The gauge boson masses are found to be

$$M_\gamma = 0, \quad M_{\mathrm{W}} = \frac{v}{2} g_2, \quad M_{\mathrm{Z}} = \frac{v}{2}\sqrt{g_1^2 + g_2^2}\,. \tag{1.20}$$

The photon is indeed massless. An important result is the relation of the ratio of the heavy gauge boson masses to the weak mixing angle:

$$\frac{M_{\mathrm{W}}}{M_{\mathrm{Z}}} = \cos\theta_{\mathrm{w}}\,. \tag{1.21}$$

This mixing angle also appears in the boson-fermion couplings which becomes more evident when the interaction term is phrased in terms of currents:

$$\mathcal{L}_{\text{int}} = -e \left\{ A_\mu J^\mu_{\text{em}} + \frac{1}{\sqrt{2}\sin\theta_{\text{w}}} \left(W^+_\mu J^\mu_{\text{CC}} + W^-_\mu J^{\mu\dagger}_{\text{CC}} \right) + \frac{1}{\sin\theta_{\text{w}}\cos\theta_{\text{w}}} Z_\mu J^\mu_{\text{NC}} \right\} . \tag{1.22}$$

The J^μ denote the electromagnetic, charged and neutral current of each fermion field ψ_f

$$J^\mu_{\text{em}} = \bar{\psi}_f \gamma^\mu (T_3 + Y) \psi_f \tag{1.23}$$

$$J^\mu_{\text{CC}} = \bar{\psi}_f \gamma^\mu (T_1 + i T_2) \psi_f \tag{1.24}$$

$$J^\mu_{\text{NC}} = \bar{\psi}_f \gamma^\mu T_3 \psi_f - \sin^2\theta_{\text{w}} J^\mu_{\text{em}} . \tag{1.25}$$

In the first term the electrical charge is identified with $Q = T_3 + Y$. The charged current term describes W boson production and decay into chiral fermions:

$$J^\mu_{\text{CC}} = \frac{1}{2} \bar{\psi}_f \gamma^\mu (1 - \gamma_5) \psi_f . \tag{1.26}$$

The neutral current is usually written in a more general way to split vector and axial-vector currents:

$$J^\mu_{\text{NC}} = \frac{1}{2} \bar{\psi}_f \left(g^{\text{f}}_{\text{V}} \gamma^\mu - g^{\text{f}}_{\text{A}} \gamma^\mu \gamma_5 \right) \psi_f , \tag{1.27}$$

with the coupling constants

$$g^{\text{f}}_{\text{V}} = T_3 - 2Q \sin^2\theta_{\text{w}} \tag{1.28}$$

$$g^{\text{f}}_{\text{A}} = T_3 , \tag{1.29}$$

and with T_3 and Q according to Table 1.1. This relates the electromagnetic coupling e to the electroweak couplings g_1 and g_2

$$e = g_1 \cos\theta_{\text{w}} = g_2 \sin\theta_{\text{w}} \tag{1.30}$$

The classical Fermi interaction of charged currents

$$\mathcal{L}_{\text{Fermi}} = -\frac{G_{\text{F}}}{\sqrt{2}} J^\dagger_{\mu,\text{CC}} J^\mu_{\text{CC}} \tag{1.31}$$

is a second order process in the Standard Model mediated by W exchange. In the limit of small momentum transfer the Fermi constant G_{F} becomes

$$G_{\mathrm{F}} = \frac{g_2^2}{4\sqrt{2}M_{\mathrm{W}}^2} \, . \tag{1.32}$$

The constant G_{F} is determined in muon lifetime measurements to the current world-average value of $G_{\mathrm{F}} = 1.16637(1) \times 10^{-5}$ GeV^{-2} [3]. The W-mass can therefore be calculated at tree level:

$$M_{\mathrm{W}} = \frac{\pi \alpha_{\mathrm{QED}}}{\sin^2 \theta_{\mathrm{w}} \sqrt{2} G_{\mathrm{F}}} \tag{1.33}$$

with the electromagnetic $\alpha_{\mathrm{QED}} = e^2/4\pi$. This implies

$$v = \frac{1}{\sqrt{\sqrt{2}G_{\mathrm{F}}}} \, . \tag{1.34}$$

which numerically is equal to $v = 246.221(2)$ GeV. All mass terms in the Standard Model are proportional to v. Apart from the gauge bosons, no other particle mass is however fixed by only this value but involves a second free parameter.

The physical manifestation of the Higgs mechanism, the neutral Higgs boson, H, interacts with the fermions and gauge bosons. The interaction Lagrangian is given by:

$$\mathcal{L}_{\mathrm{H,int}} = -\frac{m_{\mathrm{f}}}{v} H \bar{\psi}_f \psi_f + \frac{M_{\mathrm{W}}^2}{v^2} W_\mu^- W_+^\mu \left(H^2 + 2vH\right) + \frac{M_{\mathrm{Z}}^2}{2v^2} Z_\mu Z^\mu \left(H^2 + 2vH\right) \tag{1.35}$$

and the lowest order Feynman diagrams are shown in Fig. 1.1. The Higgs couplings to fermions and vector bosons, Hff, HVV, and $HHVV$, depend directly on the particle masses:

$$g_{Hff} = i\frac{m_{\mathrm{f}}}{v} \; ; \; g_{HVV} = -2i\frac{M_{\mathrm{V}}^2}{v} \; ; \; g_{HHVV} = -2i\frac{M_{\mathrm{V}}^2}{v^2} \, . \tag{1.36}$$

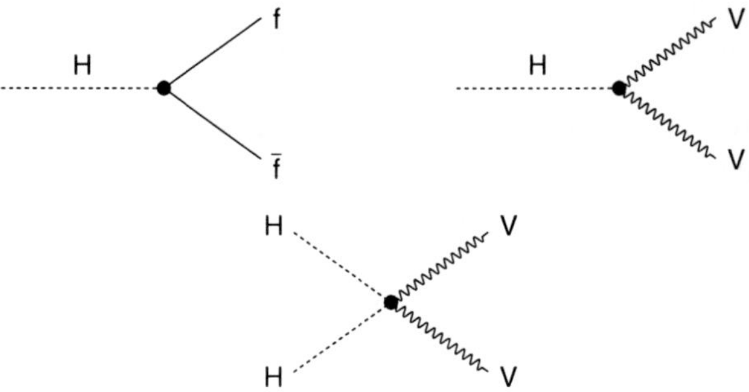

Fig. 1.1 Lowest order Feynman diagrams of Higgs boson couplings to fermions, f, and massive gauge bosons, V = W, Z

The Higgs boson has not been discovered experimentally, yet. The coupling structure is therefore important in the corresponding searches, since it is expected to couple to the heaviest particles that is kinematically allowed. Important production and decay channels with massless particles, like gluons and photons, are possible via loops, which is described later in more detail.

Indications about the energy scale of the mass of the Higgs boson can be derived from several arguments. The unitarity bound on longitudinal gauge boson scattering [9] requires new physics in the TeV energy scale. If it is the Higgs boson that dresses the scattering amplitude to not exceed the unitarity limit, the mass of the Higgs boson should not exceed 870 GeV (at tree level and in the high-energy limit [10]). Furthermore, the Higgs width into vector bosons increases to lowest order with M_{H}^3/v^2. The particle character of the Higgs boson requires that the width should not exceed M_{H}, which limits M_{H} to about 1.4 TeV [10].

Quantum effects of the Higgs self-interaction potential

$$\mathcal{L}_{\mathrm{H,self}} = \lambda v H^3 + \frac{\lambda}{4} H^4 \tag{1.37}$$

lead to additional theoretical constraints. The renormalisation group equation for the Higgs self-coupling λ behaves to first order and in the limit of large momentum transfer, $Q^2 \gg 0$, according to [11]

$$\lambda(Q^2) = \lambda(v) \frac{1}{1 - \frac{3}{4\pi^2}\lambda(v^2) \log \frac{Q^2}{v^2}} \,. \tag{1.38}$$

This means that the coupling has a Landau pole at $\Lambda_C = v e^{\frac{4\pi^2 v^2}{3M_{\mathrm{H}}^2}}$, where it becomes infinite. This typical behaviour of a ϕ^4 theory shows that it is only an effective theory up to the scale Λ_C. Thus, if Λ_C is set to the very high energies of the "grand unification" (GUT) scale of about 10^{16} GeV, the Higgs mass must not exceed ≈ 250 GeV for the theory to remain valid. This triviality bound is shown graphically in Fig. 1.2. One must however keep in mind that in case of large values of λ, perturbation theory will break down. On the other hand, lattice calculations show that this limit still stays in the range of $M_{\mathrm{H}} < 710$ GeV [11].

A lower limit on M_{H} is derived from the stability of the Higgs potential [5]. Quantum corrections to $HH \rightarrow HH$ scattering with fermion and vector boson loops tend to push $\lambda(Q^2)$ to negative values. In the small coupling limit, one obtains:

$$\lambda(Q^2) = \lambda(v^2) + \frac{1}{16\pi^2}\left\{-12\frac{m_{\mathrm{t}}^4}{v^4} + \frac{3}{16}\left(2g_2^4 + \left(g_2^2 + g_1^2\right)^2\right)\right\} \log \frac{Q^2}{v^2} \tag{1.39}$$

which can become negative if $\lambda(v^2)$ is small. The Higgs potential then develops a new minimum $V(|Q|) < V(v)$, which is not stable. To avoid the instability, the Higgs mass should fulfil

$$M_H^2 = 2\lambda(v^2)v^2 > -\frac{v^2}{8\pi^2}\left\{-12\frac{m_t^4}{v^4} + \frac{3}{16}\left(2g_2^4 + \left(g_2^2 + g_1^2\right)^2\right)\right\}\log\frac{Q^2}{v^2}, \quad (1.40)$$

which depends mainly on the top quark mass, m_t, and the values of the gauge couplings, g_i. This yields $M_H > 370$ GeV for $\Lambda_C = 10^{16}$ GeV. A more accurate calculation [12, 13] results in $M_H > 125$ GeV, depicted as vacuum stability bound in Fig. 1.2.

However, the vacuum could also be meta-stable and the electroweak minimum may differ from the absolute minimum of the effective theory. To avoid significant tunnelling probability between the two vacua, the Higgs mass should also not exceed some minimal value, which is about 10–15 GeV lower than the normal stability bound [13].

The fermion and boson loop corrections to the Higgs propagator relate the physical Higgs mass, M_H, to the "bare" mass, M_H^0, of the unrenormalised Lagrangian. The corresponding lowest order diagrams are shown in Fig. 1.3. Cutting the loop integral momenta at a scale Λ one obtains in the limit of a large top quark mass at lowest order [5]:

$$M_H^2 = \left(M_H^0\right)^2 + \frac{3\Lambda^2}{8\pi^2 v^2}\left[M_H^2 + 2M_W^2 + M_Z^2 - 4m_t^2\right]. \quad (1.41)$$

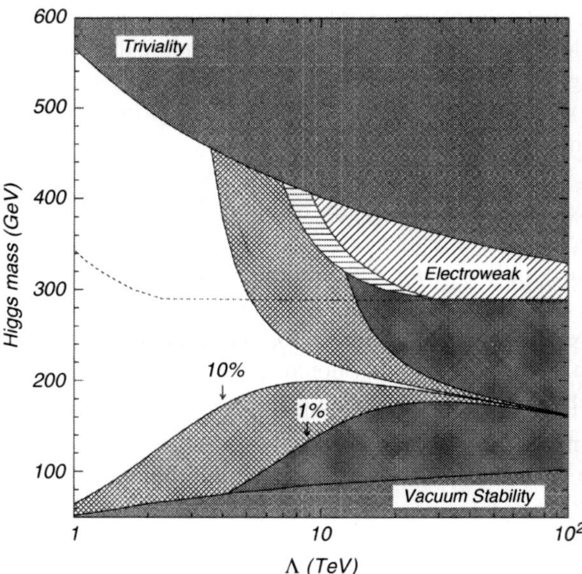

Fig. 1.2 The theoretical bounds [14] on M_H from vacuum stability (*lower bounds*) and triviality (*upper bound*) are shown as *grey areas*. They indicate the limiting values of M_H between which the Standard Model remains valid up to the energy scale Λ. The hatched regions indicate where fine-tuning at the level of 1 and 10% is necessary. The *white region* corresponds to the parameter range where all constraints are fulfilled without much fine-tuning (> 10%). The analysis of electroweak data leads to further constrains on M_H, indicated by the *dashed area*

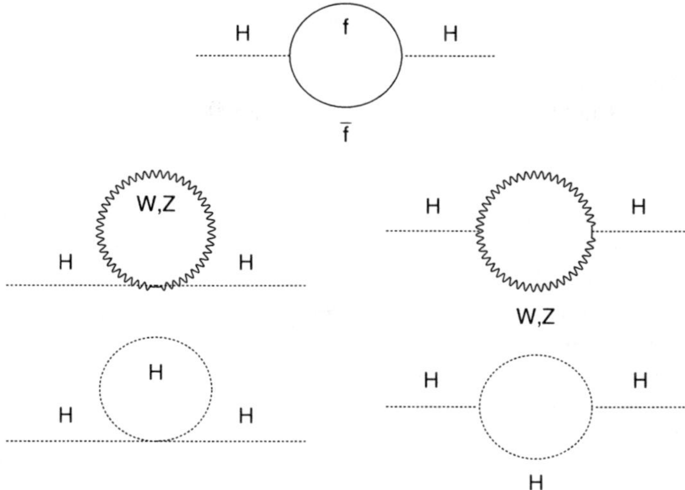

Fig. 1.3 Feynman diagrams of the one-loop corrections to the Standard Model Higgs boson mass

Now, if Λ is chosen to be at large energy scales of 10^{16} GeV, the parameter M_{H}^{0} must be tuned properly to 16 digits to get M_{H} right. Another solution would be to avoid the quadratic divergence by choosing $M_{\mathrm{H}}^{2} = -\left(2M_{\mathrm{W}}^{2} + M_{\mathrm{Z}}^{2} - 4m_{\mathrm{t}}^{2}\right)$ [15]. This is however not valid at higher orders, and the fine-tuning problem remains. Figure 1.2 shows how much fine-tuning is needed assuming the validity of the Standard Model up to a given scale Λ. Only for $M_{\mathrm{H}} \approx 200$ GeV, fine-tuning is in a reasonable range ($> 10\%$) also at high energies. This is the most stringent theoretical constraint within the Standard Model as a perturbative theory.

Eventually, also cosmological arguments which involve the formation of large scale structures of the universe due to the so-called inflation model [16] can constrain the shape of the Higgs potential. Assuming that it is the Standard Model Higgs field that initiates inflation and with certain conditions on the Higgs gravitational coupling [17], a limit of $M_{\mathrm{H}} \in [126, 194]$ GeV can be derived.

In summary, the rather general unitary bound requires new physics at the 1 TeV scale, which in the Standard Model should appear in form of the scalar Higgs boson. Further theoretical constraints indicate that $M_{\mathrm{H}} \approx 200$ GeV if the Standard Model shall remain valid to very high energy scales.

1.2 Quantum Chromodynamics

Strong interactions of quark fields ψ_q and gluon fields G_μ are described in the Standard Model Lagrangian by the following term:

$$\mathcal{L}_{\mathrm{colour}} = -\frac{1}{4} F_a^{\mu\nu} F_{\mu\nu}^a + \sum_q i\bar{\psi}_{q,j}\gamma^\mu \left(\partial_\mu - i g_3 G_\mu^a \frac{\lambda_a}{2}\right) \psi_{q,k} \,. \tag{1.42}$$

where $a = 1, \ldots, 8$ and $j, k = 1, 2, 3$ denote the colour indices for gluons and quarks, respectively. The sum extends over all quarks *u,d,s,c,t,b*. The gauge field strength of the gluon fields G_μ is given by

$$F^a_{\mu\nu} = \partial_\mu G^a_\nu - \partial_\nu G^a_\mu - g_3 f^{abc} G^b_\mu G^c_\nu \,. \tag{1.43}$$

The constant g_3 is the coupling parameter and the factors f^{abc} are the $SU(3)$ structure constants. The λ^a matrices denote the three-dimensional representation of the group generators of the $SU(3)$.

Since quark-gluon and gluon-gluon interactions are proportional to g_3^2 one introduces the strong coupling constant $\alpha_s = g_3^2/4\pi$. The most interesting property of QCD is the behaviour of α_s: when virtual corrections due to the gluon field are taken into account, the strong coupling changes with momentum transfer q^2 like:

$$\alpha_s(q^2) = \frac{12\pi}{(33 - 2n_f)\log\left(q^2/\Lambda^2_{\mathrm{QCD}}\right)} \tag{1.44}$$

with the QCD energy scale Λ_{QCD} and the number of quark flavours n_f with quark masses lower than $\sqrt{q^2}$. This means that the value of α_s decreases with increasing q^2. This effect is known as asymptotic freedom. However, the opposite behaviour is seen with decreasing q^2: the coupling strength increases. This has the consequence that no free coloured objects are observed in nature, and quarks and gluons are bound by the principle of colour confinement. The running of α_s is nicely confirmed in measurements which are compiled in [18].

The predictions of perturbative QCD are successfully applied when quarks and gluons can be considered as free particles, which is usually the case in the high energy regime where effects of colour confinement can be neglected. The transition from coloured quarks and gluons to the colourless hadronic particles in the final state of a physics reaction is however difficult to describe from first principles. In theoretical calculations, Monte Carlo models are an effective approach to cover the fragmentation and hadronisation phase of the physics process. The most common models are described at the end of this chapter.

1.3 Electroweak Radiative Corrections

Higher-order radiative corrections need to be taken into account for the theoretical calculations to match the precision of the measurements. They also lead to more involved relations between the Standard Model parameters, which are

- the fermion masses, m_f
- the electroweak boson masses, M_W, M_Z
- the mass of the Higgs boson, M_H
- the electromagnetic and strong coupling constants, α_{QED} and α_s
- the elements of the CKM mixing matrix and, in an extension to the Standard Model, those of the neutrino mixing matrix.

The coupling constants that appear in the charged and neutral current interactions are in principle functions of these parameters and of the quantum numbers of the interacting particles. The non-trivial relations between coupling and mass measurements are tested in a combined analysis to accept or reject the theoretical model.

The tree-level Eq. (1.33), which relates the W boson mass to the Z boson mass, is modified in the following way:

$$M_{\mathrm{W}}^2\left(1-\frac{M_{\mathrm{W}}^2}{M_{\mathrm{Z}}^2}\right)=\frac{\pi\alpha_{\mathrm{QED}}}{\sqrt{2}G_{\mathrm{F}}}\times\frac{1}{1-\Delta r} \tag{1.45}$$

The Δr term is due to propagator corrections caused by loop diagrams, as shown in Fig. 1.4. They can by split into QED corrections, $\Delta\alpha_{\mathrm{QED}}$, to the photon propagator, electroweak corrections, $\Delta\rho$, and an electroweak remainder term, $\Delta r_{\mathrm{remainder}}$:

$$\Delta r=\Delta\alpha_{\mathrm{QED}}-\frac{\cos^2\theta_{\mathrm{w}}}{\sin^2\theta_{\mathrm{w}}}\Delta\rho+\Delta r_{\mathrm{remainder}} \tag{1.46}$$

The QED corrections are related to the photon self-energy which change the electromagnetic coupling for non-zero momentum transfer q^2:

$$\alpha_{\mathrm{QED}}(q^2)=\alpha_{\mathrm{QED}}(0)\frac{1}{1-\Delta\alpha_{\mathrm{QED}}(q^2)} \tag{1.47}$$

where $\alpha_{\mathrm{QED}}(0)=1/137.035999679(94)$ [19]. The most interesting value is the correction at the Z-pole, $q^2=M_{\mathrm{Z}}^2$, because many precision measurements are carried out at this centre-of-mass energy.

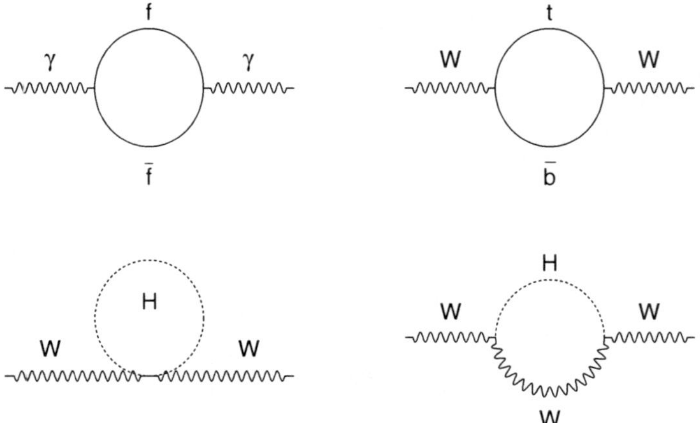

Fig. 1.4 Feynman diagrams showing leading order loop corrections to the vector boson propagators

Each fermion with mass $m < \sqrt{q^2}$ contributes to $\Delta\alpha(M_Z^2)$. The leptonic correction has been calculated to third order to be $\Delta\alpha_\ell = 0.03150$ [20] with negligible uncertainty. The corrections due to quark loops require a more detailed analysis because there are potentially large QCD corrections to be taken into account. The top quark term, $\Delta\alpha_{\rm top}$, is treated separately since it depends on the top mass. Its value is $\Delta\alpha_{\rm top} = -0.00007(1)$ [21]. The light quark term, $\Delta\alpha_{\rm had}^{(5)}$, is usually calculated from measurements of the hadronic cross-section in e^+e^- collisions at centre-of-mass energies, $\sqrt{s}$, well below the Z pole, $\sqrt{s} \ll M_Z$. An experimentally driven evaluation yields $\Delta\alpha_{\rm had}^{(5)} = 0.02758 \pm 0.0035$ [22]. This correction gives the largest uncertainty to $\Delta\alpha(M_Z^2)$.

The ρ parameter is defined as

$$\rho = \frac{1}{\cos^2\theta_{\rm w}} \times \frac{M_{\rm W}^2}{M_{\rm Z}^2} \tag{1.48}$$

and has a value of 1 at tree level. The quantum corrections to this relation is mainly determined by the self-energy of the W boson propagator. It is sensitive to all $SU(2)$ multiplets which directly couple to gauge bosons and exhibit a large mass splitting. The mass differences in the light quark multiplets are in general small. The leading term is therefore given by the $t-b$ loop:

$$\Delta\rho_t = 3\frac{G_{\rm F} m_{\rm t}^2}{8\pi^2\sqrt{2}} = 0.00939 \pm 0.00014 \tag{1.49}$$

in the approximation $m_{\rm t} \gg m_b$, and using the recent measurement of the top quark mass, $m_{\rm t} = 173.1 \pm 1.3$ GeV [23], by the CDF and DØ collaborations.

Higgs boson contributions to $\Delta\rho$ are playing an interesting role in the analysis of measurements in the framework of the Standard Model. The corrections are

$$\Delta\rho_H = -3\frac{G_{\rm F} M_{\rm W}^2}{8\pi^2\sqrt{2}} \tan^2\theta_{\rm w} \left(\log\frac{M_{\rm H}^2}{M_{\rm W}^2} - \frac{5}{6}\right) \tag{1.50}$$

for $M_{\rm H} \gg M_{\rm W}$.

Also in the remainder term Higgs and top quark contributions appear:

$$\begin{aligned}\Delta r_{t,\rm rem} &= -\frac{G_{\rm F} M_{\rm W}^2}{8\pi^2\sqrt{2}} \left\{ 3\cot^2\theta_{\rm w}\frac{m_{\rm t}^2}{M_{\rm W}^2} + 2\left(\cot^2\theta_{\rm w} - \frac{1}{3}\right)\log\frac{m_{\rm t}^2}{M_{\rm W}^2} \right. \\ &\quad \left. + \frac{4}{3}\log\cos^2\theta_{\rm w} + \cot^2\theta_{\rm w} - \frac{7}{9} \right\} \qquad (1.51) \\ \Delta r_{H,\rm rem} &= \frac{\sqrt{2} G_{\rm F} M_{\rm W}^2}{16\pi^2} \left\{ \frac{11}{3}\left(\log\frac{M_{\rm H}^2}{M_{\rm W}^2} - \frac{5}{6}\right) \right\}, \qquad (1.52)\end{aligned}$$

again for $M_{\rm H} \gg M_{\rm W}$.

Radiative corrections are as well important in the determination of the Z couplings to the fermions, g_V^f and g_A^f, and their effective values are defined as

$$g_\mathrm{V}^\mathrm{f,eff} = \sqrt{\rho_\mathrm{f}}\left(T_3 - 2Q_\mathrm{f}\sin^2\theta_\mathrm{eff}^\mathrm{f}\right) \tag{1.53}$$

$$g_\mathrm{A}^\mathrm{f,eff} = \sqrt{\rho_\mathrm{f}}T_3 \tag{1.54}$$

with an effective weak mixing angle, $\sin^2\theta_\mathrm{eff}^\mathrm{f}$, and the ρ_f parameter which includes universal Z propagator and flavour specific vertex corrections. This eliminates the dependency of the measurements on radiative corrections and reduces the measurement uncertainty. Using this definition, the ratio of the effective vector and axial-vector coupling is directly related to the effective weak mixing angle, given by

$$\sin^2\theta_\mathrm{eff} = \frac{1}{4}\left(1 - \frac{g_\mathrm{V}^\mathrm{eff}}{g_\mathrm{A}^\mathrm{eff}}\right) \tag{1.55}$$

There is an effective angle for each type of fermion, which is proportional to the on-shell definition of the mixing angle (see Eq. (1.21)):

$$\sin^2\theta_\mathrm{eff}^\mathrm{f} = \kappa_\mathrm{f}\sin^2\theta_\mathrm{w}\,. \tag{1.56}$$

The factor κ_f is related to the radiative correction term Δr_f by the following equation:

$$\sqrt{2}G_\mathrm{F}M_\mathrm{Z}^2\sin^2\theta_\mathrm{eff}^\mathrm{f}\cos^2\theta_\mathrm{eff}^\mathrm{f} = \frac{\pi\alpha_\mathrm{QED}}{1-\Delta r_\mathrm{f}}\,. \tag{1.57}$$

The quantity Δr_f is very similar to the one that is given in Eq. (1.46):

$$\Delta r_\mathrm{f} = \Delta\alpha_\mathrm{QED} - \frac{\cos^2\theta_\mathrm{w}}{\sin^2\theta_\mathrm{w}}\Delta\rho + \Delta r_\mathrm{f,rem}\,. \tag{1.58}$$

Only the last term $\Delta r_\mathrm{f,rem}$ is defined differently and takes additional $\mathrm{Z}/\gamma\rightarrow\mathrm{f\bar{f}}$ vertex corrections into account. A more detailed discussion can be found in [24].

Current calculations include electroweak radiative corrections at two-loop order to the W boson propagator. Complete fermionic two-loop results are available for the determination of $\sin^2\theta_\mathrm{eff}^\ell$. In the limit of large m_t the top contributions to $\Delta\rho$ are known to three-loop order.

Since precision measurements are well sensitive to these small quantum corrections there is sensitivity to the mass of the Higgs boson, which is the only particle of the Standard Model that has not been observed, yet. Other indirect determinations of Standard Model parameters work out well. The derived mass of the W boson $M_\mathrm{W} = 80.364 \pm 0.020$ GeV agrees well with the direct measurement at LEP and Tevatron of $M_\mathrm{W} = 80.399 \pm 0.023$ GeV. The indirect top quark mass $m_\mathrm{t} = 179.3^{+11.6}_{-8.5}$ GeV [26] has a much lower precision than the direct measurement by CDF and DØ, $m_\mathrm{t} = 173.1 \pm 1.3$ GeV [23], but also here the agreement is very

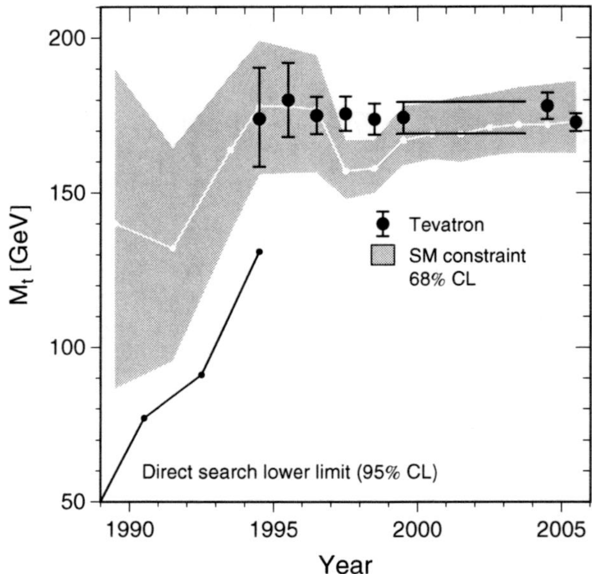

Fig. 1.5 Historic development of the indirect limits on the top mass, the direct search limits and eventually the measurements at the Tevatron [25]

good. In fact, the discovery of the top quark was lead by more and more precise model predictions. The historical development of these calculations and the first measurements are compared in Fig. 1.5. This gives confidence that also the indirect information on M_{H} is useful within the Standard Model framework.

From the analysis of electroweak precision measurements alone, an upper limit of 157 GeV at 95% confidence level (C.L.) can be derived [26]. Combined with direct searches for the Higgs boson at LEP [27], this constraint is weakened slightly and a 95% C.L. range of 114.4 GeV $< M_{\mathrm{H}} <$ 186 GeV for the mass of the Higgs boson is determined [26]. This is well in the reach of the LHC and Higgs boson searches and dedicated analyses concentrate on the low M_{H} region.

1.4 Extensions to the Standard Model

The most attractive theoretical extension to the Standard Model is super-symmetry (SUSY), which is introducing a global symmetry between bosons and fermions by changing the spin by $\pm 1/2$ units. The corresponding operators, Q_α, transform fermions into bosons and vice-versa:

$$Q|\text{Fermion}\rangle = |\text{Boson}\rangle \; ; \quad Q|\text{Boson}\rangle = |\text{Fermion}\rangle \tag{1.59}$$

They are spinors and follow the SUSY algebra:

$$\{Q_\alpha, \bar{Q}_\beta\} = -(g^\mu)_{\alpha\beta} P_\mu \; , \tag{1.60}$$

where P_μ is the momentum generator of space-time translations. The particle spectrum of the Standard Model is preserved and extended by super-symmetric partners of the known elementary particles. The partners of fermions are scalar sfermions, and the gauge boson sector is mapped to spin-1/2 gauginos. Super-symmetric models overcome some of the deficits of the Standard Model. If SUSY is exact, the fine-tuning problem is resolved due to opposite-sign loop contributions from fermions and bosons. However, the symmetry can evidently not be exact and is broken at some SUSY energy scale, since the not yet discovered SUSY partners must be of larger mass than the currently known Standard Model particles.

In super-symmetric models the Higgs-sector is necessarily extended to two Higgs doublets to avoid anomalies and to provide super-symmetric mass terms for up- and down-type fermions. The first doublet, H_1, is giving masses to the down-type fermions and the second, H_2, introduces masses to the up-type fermions. This results in three neutral Higgs bosons, h^0, H^0, A^0, and one charged Higgs boson, $H^\pm$. The neutral Higgs fields h^0 and H^0 are CP even, while the A^0 field is CP odd. The super-partners of the weak gauge bosons and the Higgs super-partners actually mix and form neutralinos, $\tilde{\chi}^0_{1,2,3,4}$, and charginos, $\tilde{\chi}^\pm_{1,2}$.

An important parameter of SUSY is the ratio of the vacuum expectation values, $\tan\beta = v_2/v_1$, of the Higgs doublet fields. Taking the splitting of the mass scale of up- and down-type fermions into account, one can argue that $\tan\beta$ should be in the order of $m_t/m_b \approx 40$. Experimental constraints will be discussed later (see Chap. 4).

SUSY particle production is usually studied within the gravity and gauge mediated minimal SUSY models Minimal Super-Gravity (mSUGRA)) [28] and Gauge-Mediated Super-symmetry Breaking (GMSB) [29]. Benchmark scenarios are chosen to cover a wide range of experimental signatures. In both models, R-parity defined as $R = (-1)^{3B+L+2S}$ with lepton number, L, baryon number, B, and spin, S, is conserved. As a consequence, SUSY particles can only be produced in pairs and the lightest SUSY particle (LSP) is stable. This leads to typical detector signatures from the SUSY decay chains since the LSP is expected to be only weakly interacting.

Typical mSUGRA models analysed at the LHC are for example, $SU1$, ..., $SU8.1$ [30], with different values of the universal sfermion and gaugino masses at the GUT scale, m_0 and $m_{1/2}$, of $\tan\beta = v_2/v_1$, of the sign of the Higgsino mass parameter, μ, and of the universal trilinear coupling, A_0, at the GUT scale. The next-to-leading order (NLO) total summed SUSY cross-section at LHC centre-of-mass energies of 14 TeV varies between 6 pb ($SU6$) and 402 pb ($SU4$) [31] for these models.

The cross-sections for SUSY Higgs production at the LHC is in the order of 1,000 pb for large $\tan\beta = 30$ and small Higgs masses of 100 GeV [5], down to 0.1 pb for large Higgs masses of 1,000 GeV. An interesting fact of SUSY models is the upper mass limit on the lightest Higgs boson, h, which is in the order of 110–130 GeV [33], depending on the mixing in the super-symmetric top sector. Because the couplings to up-type fermions are enhanced for $\tan\beta > 1$, the largest branching fraction of the h boson are to b-quark and τ lepton pairs ($\approx 90\%$ and $\approx$

10%, respectively). These are also the main decay channels of heavy CP-even Higgs bosons for large $\tan\beta$. For smaller values of $\tan\beta$ and 125 GeV $< M_{\mathrm{H}} <$ 250 GeV, the decays of the H boson are similar to the Standard Model Higgs boson, while for higher masses also the decay channels to h boson and top quark pairs open up. The CP-odd Higgs boson decays predominantly to $\mathrm{b\bar{b}}$ and $\tau^+\tau^-$ and for high masses to top quark pairs. The charged Higgs decays mainly to $\tau\nu$ for masses up to $M_{\mathrm{H}^\pm} \approx m_{\mathrm{t}}$, above which the tb final state is preferred. These final states are therefore the main search channels for super-symmetric Higgs bosons. The mass spectrum of the Higgs bosons depends mainly on the mass of the CP-odd Higgs boson, M_A, which is shown in Fig. 1.6, together with the different Higgs bosons that can be discovered at the LHC in different regions of the SUSY parameter space [32, 30].

In this work, however, only general aspects of SUSY will be discussed in the framework of the precision electroweak measurements. Further details can be found in [5, 30, 34, 35].

The Higgs sector may be enriched by adding more or higher Higgs multiplets. Such models are all constrained by the fact that the ρ parameter should not deviate too much from the measured value of 1. For a set of Higgs bosons with vacuum expectation values v_i, isospin I_i and third component I_i^3, the tree-level value of ρ is given by:

$$\rho = \frac{\sum_i \left\{ I_i(I_i+1) - \left(I_i^3\right)^2 \right\} v_i^2}{2\sum_i \left(I_i^3\right)^2 v_i^2} \,. \tag{1.61}$$

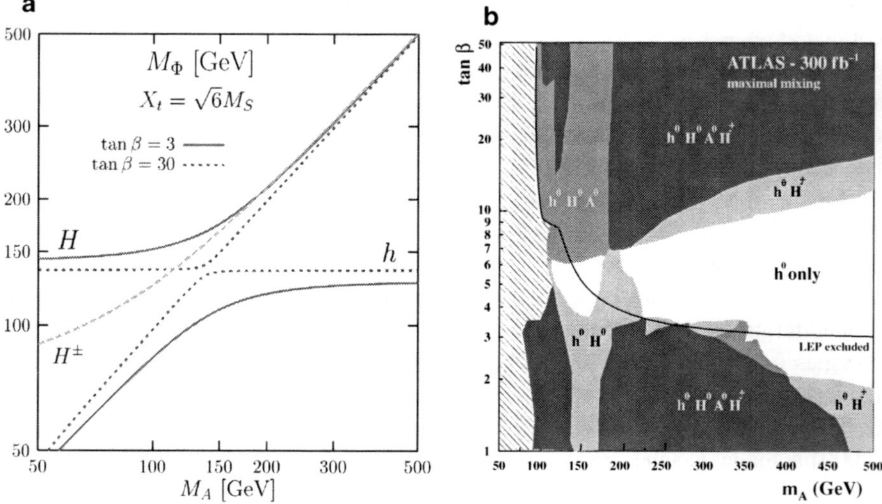

Fig. 1.6 (**a**) Super-symmetric Higgs boson masses as a function of the mass parameter, M_A, for the maximal top/stop mixing scenario [10]. (**b**) The number of SUSY Higgs bosons which can be discovered by the ATLAS experiment for different regions of the $M_A - \tan\beta$ plane assuming an integrated luminosity of 300 fb^{-1} [32]

For the Standard Model and multi-Higgs-doublet models, like SUSY, one always obtains a value of 1. If up- and down-type fermions would be of equal mass in each generation, $\rho = 1$ would hold exactly even at higher orders in the Standard Model since a larger $SU(2)_L \times SU(2)_R$ symmetry would be apparent. The top/bottom mass splitting is however breaking this symmetry. Higgs triplet models are possible extensions of the Standard Model and provide a mechanism for $SU(2)_L \times SU(2)_R$ symmetry breaking [36]. These models are also attractive because it is possible to construct neutrino mass terms compatible with current observations. One way to explain the smallness of the neutrino masses is the so-called see-saw mechanism which mixes right- and left-handed neutrinos such that heavy and light mass eigenstates evolve [37]. Mass terms of this kind can for example be constructed in Higgs triplet models. Experimentally, one should observe in addition to a rather light neutral Higgs boson [38], single- and double-charged Higgs boson, which are however not found, yet [39].

Higgs triplets are also predicted in Little Higgs models [40] in which the Higgs sector is dynamically generated by the interaction of originally massless scalar fields. The Higgs is therefore a composite particle. However, new massive vector bosons and fermions as well as additional heavy up-type quarks are predicted in the model, which have not yet been seen in experiments.

Although the Higgs mechanism is very attractive for breaking electroweak symmetry and providing particle masses, there may be alternatives [41] which explain these phenomena without a Higgs field. The strong $W_L W_L$ scattering must then be unitarized by some other states, e.g., techni-ρ particles in Technicolour models [42] or Kaluza-Klein (KK) gauge bosons in Higgs-less Models [43]. A generalised treatment of these models can be performed in terms of an effective Lagrangian method [44], which allows the study of possible effects in electroweak boson scattering and signatures at the LHC. New resonance states as well as anomalous scattering cross-sections can be expected in these scenarios.

A requirement of all more or less exotic extensions of the Standard Model is the necessity to be compatible with todays precise measurements in the electroweak sector. In the remaining part of this chapter the phenomenology of the most important Standard Model processes will therefore be discussed.

1.5 Z Boson Production and Decay in e^+e^- Collisions

The properties of the Z boson were studied in detail at LEP and SLD at energies around the Z pole. The following paragraphs summarise the most important observables that enter into the global analysis of electroweak data. Further details can be found in [45].

At energies below the W-pair threshold, Z bosons only decay to fermion pairs. The diagrams that contribute to the process $e^+e^- \to f\bar{f}$ at lowest order are shown in Fig. 1.7. The differential cross-section with photon and Z exchange, as well as their interference, can be written in the following way

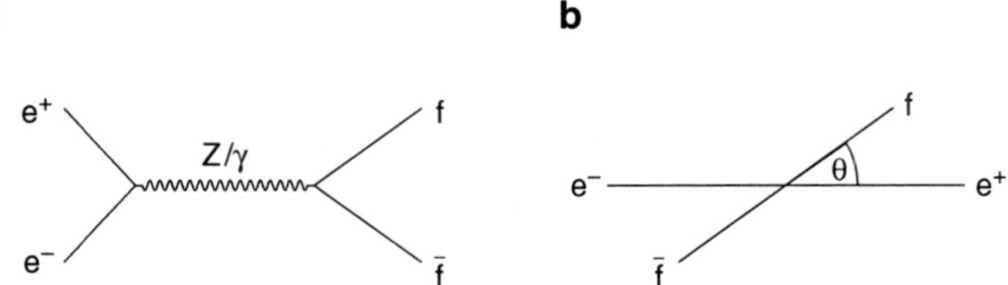

Fig. 1.7 (**a**) Feynman diagrams for fermion-pair production in e^+e^- collisions with photon and Z-boson exchange. (**b**) The scattering angle, θ, between incoming electron and final state fermion

$$\begin{aligned}\frac{d\sigma}{d\cos\theta}(\mathrm{e^+e^-}\to \mathrm{f\bar{f}}) = \frac{\pi N_c^{\mathrm{f}}}{2s}\Big\{&|\alpha(s)Q_{\mathrm{f}}|^2(1+\cos^2\theta)\\ &-8Re\left(\alpha^* Q_{\mathrm{f}}\chi(s)\left[\mathcal{G}_{\mathrm{Ve}}\mathcal{G}_{\mathrm{Vf}}(1+\cos^2\theta)+2\mathcal{G}_{\mathrm{Ae}}\mathcal{G}_{\mathrm{Af}}\cos\theta\right]\right)\\ &+16|\chi(s)|^2\left[(|\mathcal{G}_{\mathrm{Ve}}|^2+|\mathcal{G}_{\mathrm{Ae}}|^2)(|\mathcal{G}_{\mathrm{Vf}}|^2+|\mathcal{G}_{\mathrm{Af}}|^2)(1+\cos^2\theta)\right.\\ &\left.+8Re\left(\mathcal{G}_{\mathrm{Ve}}\mathcal{G}_{\mathrm{Ae}}^*\right)Re\left(\mathcal{G}_{\mathrm{Vf}}\mathcal{G}_{\mathrm{Af}}^*\right)\cos\theta\right]\Big\} \qquad (1.62)\end{aligned}$$

where complex coupling constants $\alpha(s)$, $\mathcal{G}_{\mathrm{Vf}}$, $\mathcal{G}_{\mathrm{Af}}$ are used to absorb electroweak corrections (see [45]). Their real parts are related to the real couplings $g_{\mathrm{V}}^{\mathrm{f}}$ and $g_{\mathrm{A}}^{\mathrm{f}}$ by

$$\frac{g_{\mathrm{V}}^{\mathrm{f}}}{g_{\mathrm{A}}^{\mathrm{f}}} = Re\frac{\mathcal{G}_{\mathrm{Vf}}}{\mathcal{G}_{\mathrm{Af}}} = 1 - 4Q_{\mathrm{f}}\sin^2\theta_{\mathrm{eff}}^{\mathrm{f}}\,. \qquad (1.63)$$

The polar angle θ is the angle between the produced fermion and the incoming electron beam, as illustrated in Fig. 1.7. The propagator term

$$\chi(s) = \frac{G_{\mathrm{F}}M_{\mathrm{Z}}^2}{8\pi\sqrt{2}}\frac{s}{s-M_{\mathrm{Z}}^2+is\Gamma_{\mathrm{Z}}/M_{\mathrm{Z}}} \qquad (1.64)$$

is defined with an s-dependent width, $\Gamma_{\mathrm{Z}} = \Gamma_{\mathrm{Z}}(s)$. This is the convention used for all mass measurements at LEP. The alternative mass definition as the real part of the complex pole corresponds to a propagator with an s-independent width:

$$\bar{\chi}(s) = \frac{G_{\mathrm{F}}\bar{M}_{\mathrm{Z}}^2}{8\pi\sqrt{2}}\frac{s}{s-\bar{M}_{\mathrm{Z}}^2+i\bar{\Gamma}_{\mathrm{Z}}\bar{M}_{\mathrm{Z}}}\,. \qquad (1.65)$$

The two sets of variables are related by

$$M_{\mathrm{Z}} = \bar{M}_{\mathrm{Z}}\sqrt{1+\bar{\Gamma}_{\mathrm{Z}}^2/\bar{M}_{\mathrm{Z}}^2} \approx \bar{M}_{\mathrm{Z}} + 34.1\ \mathrm{MeV} \qquad (1.66)$$

$$\bar{\Gamma}_{\mathrm{Z}} = \bar{\Gamma}_{\mathrm{Z}}\sqrt{1+\bar{\Gamma}_{\mathrm{Z}}^2/\bar{M}_{\mathrm{Z}}^2} \approx \bar{\Gamma}_{\mathrm{Z}} + 0.9\ \mathrm{MeV} \qquad (1.67)$$

Both propagators lead to the same resonance shape $\sigma(s)$. For numerical results the s-dependent width scheme is used.

The partial width of the Z decaying into fermion pairs is given by

$$\Gamma_{f\bar{f}} = N_c^f \frac{G_F M_Z^3}{6\sqrt{2}\pi} \left(|\mathcal{G}_{Af}|^2 R_{Af} + |\mathcal{G}_{Vf}|^2 R_{Vf} \right) + \Delta_{ew.QCD} \tag{1.68}$$

The radiator factors R_{Vf} and R_{Af} take into account final state QED and QCD corrections, while by $\Delta_{ew,QCD}$ small contributions from non-factorisable corrections are included. To first order the R factors for axial and vector coupling are equal and given by

$$R_{Vf} = R_{Af} = R_f = R_{QED} R_{QCD} \tag{1.69}$$

with QED correction terms for all charged fermions

$$R_{QED} = 1 + \frac{3}{4} Q_f^2 \frac{\alpha(M_Z^2)}{\pi} + \dots \tag{1.70}$$

and QCD correction for quarks

$$R_{QCD} = 1 + \frac{\alpha_s(M_Z^2)}{\pi} + \dots . \tag{1.71}$$

This inclusive definition of the fermionic decay width with quantum corrections simplifies the relation to the total decay width of the Z boson:

$$\Gamma_Z = \Gamma_{ee} + \Gamma_{\mu\mu} + \Gamma_{\tau\tau} + \Gamma_{had} + \Gamma_{inv} \tag{1.72}$$

where the hadronic width is the sum of the quark decay widths

$$\Gamma_{had} = \sum_{q \neq t} \Gamma_{q\bar{q}} . \tag{1.73}$$

The so-called invisible width sums up the contributions from neutrino decays

$$\Gamma_{inv} = N_\nu \Gamma_{\nu\bar{\nu}} . \tag{1.74}$$

The factor N_ν is the number of light neutrino generations and equal to 3 in the Standard Model. By measuring total and partial width of the Z boson this identity can be verified experimentally.

Furthermore, the total cross-section of the $\cos\theta$-symmetric Z production term is written as

$$\sigma_{f\bar{f}}^Z = \frac{\sigma_{f\bar{f}}^0}{R_{QED}} \frac{s\Gamma_Z^2}{\left(s - M_Z^2\right)^2 + s^2\Gamma_Z^2/M_Z^2} \tag{1.75}$$

with the pole cross-section

$$\sigma^0_{\mathrm{f\bar{f}}} = \frac{12\pi}{M_Z^2}\frac{\Gamma_{\mathrm{ee}}\Gamma_{\mathrm{f\bar{f}}}}{\Gamma_Z^2} . \tag{1.76}$$

The hadronic pole cross-section σ^0_{had} and the hadronic to leptonic branching ratios

$$R^0_\ell = \frac{\Gamma_{\mathrm{had}}}{\Gamma_{\ell\ell}} \tag{1.77}$$

are observables that are included in the global analysis of electroweak data.

Since the axial couplings between Z and fermion, g_A^f, are non-zero, there is an asymmetry in the number of events with the fermion produced in the forward ($\theta > \pi/2$) and backward ($\theta < \pi/2$) hemispheres. This asymmetry is experimentally determined as

$$A_{\mathrm{FB}} = \frac{N_F - N_B}{N_F + N_B} , \tag{1.78}$$

where N_F is the number of events with a forward scattered fermion, and N_B the number of events with a backward scattered fermion. If only Z boson exchange is assumed the differential cross-section is simplified. For the case of incoming polarised electrons and unpolarised positrons, and averaged over final state helicities it is given by:

$$\frac{d\sigma_{\mathrm{f\bar{f}}}}{d\cos\theta} = \frac{3}{8}\sigma^{\mathrm{tot}}_{\mathrm{f\bar{f}}}\left[(1-\mathcal{P}_e\mathcal{A}_e)(1+\cos^2\theta) + 2(\mathcal{A}_e - \mathcal{P}_e)\mathcal{A}_f\cos\theta\right] , \tag{1.79}$$

with the electron polarisation $\mathcal{P}_e$ and the asymmetry parameter

$$\mathcal{A}_f = \frac{2g_V^f g_A^f}{\left(g_V^f\right)^2 + \left(g_A^f\right)^2} \tag{1.80}$$

The forward-backward asymmetry is therefore equal to

$$A_{\mathrm{FB}} = \frac{\int_0^{+1}\frac{d\sigma\mathrm{f\bar{f}}}{d\cos\theta}d\cos\theta - \int_{-1}^{0}\frac{d\sigma\mathrm{f\bar{f}}}{d\cos\theta}d\cos\theta}{\int_{-1}^{+1}\frac{d\sigma\mathrm{f\bar{f}}}{d\cos\theta}d\cos\theta} = \frac{\sigma_F - \sigma_B}{\sigma_F + \sigma_B} = \frac{3}{4}\mathcal{A}_e\mathcal{A}_f . \tag{1.81}$$

When the beam is polarised, the left-right asymmetry can be determined by measuring the event rate difference for positive and negative polarisation $\mathcal{P}_e$ of the incoming electrons:

$$A_{\mathrm{LR}} = \frac{\sigma_L - \sigma_R}{\sigma_L + \sigma_R} = \mathcal{A}_e . \tag{1.82}$$

And finally, the left-right forward-backward asymmetry is proportional to $\mathcal{A}_\mathrm{f}$:

$$A_{\mathrm{LR,FB}} = \frac{(\sigma_F - \sigma_B)_L - (\sigma_F - \sigma_B)_R}{(\sigma_F + \sigma_B)_L + (\sigma_F + \sigma_B)_R} \frac{1}{\langle|\mathcal{P}_\mathrm{e}|\rangle} = \frac{3}{4}\mathcal{A}_\mathrm{f} \tag{1.83}$$

In case of the tau lepton, also the final state helicity can be measured. The correspond polarisation is defined as

$$\mathcal{P}_\mathrm{f}(\cos\theta) = \frac{d(\sigma_r - \sigma_l)}{d\cos\theta}\left(\frac{d(\sigma_r + \sigma_l)}{d\cos\theta}\right)^{-1} \tag{1.84}$$

and given by

$$\mathcal{P}_\mathrm{f}(\cos\theta) = -\frac{\mathcal{A}_\mathrm{f}(1+\cos^2\theta) + 2\mathcal{A}_\mathrm{e}\cos\theta}{(1+\cos^2\theta) + 2\mathcal{A}_\mathrm{e}\mathcal{A}_\mathrm{f}\cos\theta} \tag{1.85}$$

and its average is

$$\langle\mathcal{P}_\mathrm{f}\rangle = -\mathcal{A}_\mathrm{f} \tag{1.86}$$

The asymmetry observables on the Z pole are defined without further radiative corrections, unlike the Z decay widths. To extract the asymmetry from data the measurements are corrected for radiative effects, γ exchange and $\gamma - \mathrm{Z}$ interference terms. The pole quantities derived in this way are eventually

$$A_{\mathrm{FB}}^{0,\mathrm{f}} = \frac{3}{4}\mathcal{A}_\mathrm{e}\mathcal{A}_\mathrm{f} \tag{1.87}$$

$$A_{\mathrm{LR}}^{0} = \mathcal{A}_\mathrm{e} \tag{1.88}$$

$$A_{\mathrm{LR,FB}}^{0} = \frac{3}{4}\mathcal{A}_\mathrm{f} \tag{1.89}$$

$$\langle\mathcal{P}_\tau^0\rangle = -\mathcal{A}_\tau \tag{1.90}$$

$$A_{\mathrm{FB}}^{\mathrm{pol},0} = -\frac{3}{4}\mathcal{A}_\mathrm{e} \tag{1.91}$$

Since the parameter $\mathcal{A}_\mathrm{e}$ is measured in left-right asymmetries independently from the forward-backward asymmetries, also the individual parameter $\mathcal{A}_\mu$, $\mathcal{A}_\tau$, $\mathcal{A}_\mathrm{b}$ and $\mathcal{A}_\mathrm{c}$ can be extracted. When expressing them as functions of the weak mixing angle one finds from Eqs. (1.80) and (1.55):

$$\begin{aligned}\mathcal{A}_\mathrm{f} &= 2\frac{g_\mathrm{V}^\mathrm{f}/g_\mathrm{A}^\mathrm{f}}{1+\left(g_\mathrm{V}^\mathrm{f}/g_\mathrm{A}^\mathrm{f}\right)^2} = 2\left(1-4|Q_\mathrm{f}|\sin^2\theta_{\mathrm{eff}}^\mathrm{f}\right)\frac{1}{\left(1+\left(1-4|Q_\mathrm{f}|\sin^2\theta_{\mathrm{eff}}^\mathrm{f}\right)^2\right)} \\ &= \frac{1-4|Q_\mathrm{f}|\sin^2\theta_{\mathrm{eff}}^\mathrm{f}}{1-4|Q_\mathrm{f}|\sin^2\theta_{\mathrm{eff}}^\mathrm{f} + 8|Q_\mathrm{f}|^2\sin^4\theta_{\mathrm{eff}}^\mathrm{f}}\,.\end{aligned} \tag{1.92}$$

This shows that the determination of the asymmetry parameters is a sensitive measurement of $\sin^2\theta^{\mathrm{f}}_{\mathrm{eff}}$, deeply connected to the symmetry breaking mechanism of the Standard Model.

1.6 W Boson Production at LEP

The production of W boson pairs in $\mathrm{e}^+\mathrm{e}^-$ collisions gives further handles for Standard Model tests. The W mass can be measured directly from the invariant mass of its decay products. Triple gauge boson couplings (TGC) of photon and Z to the W bosons as well as quartic couplings (QGC) can be determined from the analysis of the production angles and the polarisation of the W's. The fraction of longitudinal to transverse polarisation of the W bosons is measured as well.

To lowest order, the production of W-pairs is described by two Feynman diagrams, the t-channel neutrino exchange and the s-channel Z/γ exchange, as shown in Fig. 1.8. The diagram with a Higgs propagator is suppressed by a factor $m_\mathrm{e}/M_\mathrm{W}$ and can be neglected. The s-channel graph involves the non-abelian gauge boson couplings, which are described later in more detail.

The lowest-order cross-section for on-shell production of W-pairs near the threshold is given by

$$\frac{d\sigma}{d\Omega} \approx \frac{\alpha^2_{\mathrm{QED}}}{s}\frac{1}{4\sin^4\theta_\mathrm{w}}\beta\left[1 + 4\beta\cos\theta\frac{3\cos^2\theta_\mathrm{w} - 1}{4\cos^2\theta_\mathrm{w} - 1} + O(\beta^2)\right] . \tag{1.93}$$

The leading term is proportional to the W velocity, β, and is from t-channel neutrino exchange. It is the dominating term in the production threshold region, where $\sqrt{s} \approx 2M_\mathrm{W}$. For the total cross section one finds

$$\sigma \approx \frac{\pi\alpha^2_{\mathrm{QED}}}{s}\frac{1}{4\sin^4\theta_\mathrm{w}}4\beta + O(\beta^3) \tag{1.94}$$

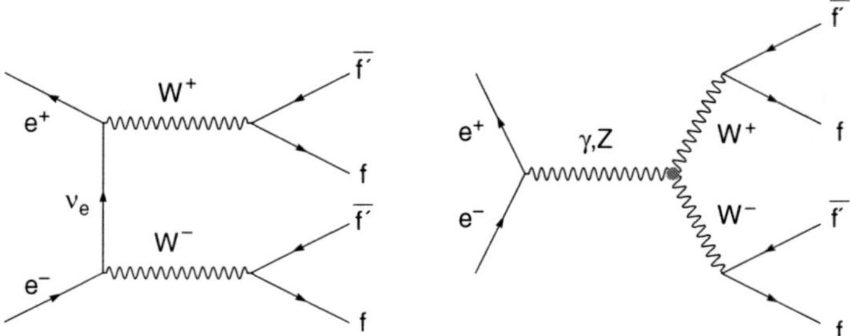

Fig. 1.8 Feynman diagrams for W-pair production at LEP with neutrino t-channel and Z/γ exchange

The terms proportional to β^2 drop out, and s-channel and interference contributions are only proportional to β^3. There is therefore no sensitivity to TGCs in the W-pair cross-section close to the threshold.

For a more complete description of the $e^+e^- \to WW$ process, also the decay of the W bosons needs to be taken into account. W bosons decay into pairs of quarks, one up-type and one down-type quark, $q\overline{q'}$, or a lepton and the corresponding neutrino, $\ell\nu_\ell$. Possible final states are therefore fully hadronic qqqq, semi-leptonic $qq\ell\nu$, or fully leptonic $\ell\nu\ell\nu$. This makes W pair production part of the so-called four-fermion processes, $e^+e^- \to ffff$. They are usually denoted as charged current, CC, and neutral current, NC, processes, depending on the boson that is exchanged in the signal process. The number of diagrams that are contributing to each final state in W-pair production is listed in Table 1.2. In this nomenclature, tree-level W-pair production is a CC03 process, while the complete description for the $qqe\nu$ final state, e.g., is of type CC20.

The CC03 cross-section takes also off-shell W-pairs into account. In a simplified form, the double-differential cross-section $\sigma_0^{CC03}(s; s+, s-) = \frac{d\sigma^{CC03}}{ds_+ ds_-}$ is folded with the Breit-Wigner propagator terms $\rho_W(s_\pm)$ [46]:

$$\sigma^{CC03}(s) = \int_0^s ds_+ \int_0^{(\sqrt{s}-\sqrt{s_+})^2} ds_- \rho_W(s_+)\rho_W(s_-)\sigma_0^{CC03}(s; s+, s-) \,, \tag{1.95}$$

where $s_+ = k_+^2$ and $s_- = k_-^2$ are the squared four-vectors of the internal W bosons. The Breit-Wigner factors are given by:

$$\rho_W(s_\pm) = \frac{1}{\pi} \frac{M_W \Gamma_W}{\left|s_\pm - M_W^2 + i M_W \Gamma_W\right|^2} \times \text{BR} \,, \tag{1.96}$$

with the branching fraction BR of the corresponding decay channel. The on-shell expression is recovered by letting the W width, Γ_W, go to zero:

$$\rho_W(s_\pm) \to \delta\left(s_\pm - M_W^2\right) \times \text{BR} \quad \text{for} \quad \Gamma_W \to 0 \tag{1.97}$$

In the simple analytic approach the main corrections from initial state photon radiation (ISR) may be included as well. Photons emitted by the incoming electrons reduce the effective centre-of-mass energy, $\sqrt{s'} < \sqrt{s}$. In the case of a single photon

Table 1.2 Number of four-fermion diagrams for the different final states in W-pair production

	$\bar{d}u$	$\bar{s}c$	$e^+\nu_e$	$\mu^+\nu_\mu$	$\tau^+\nu_\tau$
$d\bar{u}$	43	11	20	10	10
$s\bar{c}$	11	43	20	10	10
$e^-\bar{\nu}_e$	20	20	56	18	18
$\mu^-\bar{\nu}_\mu$	10	10	18	19	9
$\tau^-\bar{\nu}_\tau$	10	10	18	9	19

of energy E_γ radiated parallel to the beam, $\sqrt{s'}$ is given by

$$\sqrt{s'} = \sqrt{s}\sqrt{1 - \frac{2E_\gamma}{\sqrt{s}}} \,. \tag{1.98}$$

Photon emission with energy fractions x_i from each beam can be described by structure functions $D(x_i, s)$ which need to be convoluted

$$F(x, s) = \int_0^1 dx_1 dx_2 \delta(x - x_1 x_2) D(x_1, s) D(x_2, s) \tag{1.99}$$

The improved cross-section including ISR is then

$$\sigma^{\mathrm{CC03,ISR}}(s) = \int_{(s'/s)_{\min}}^{(s'/s)_{\max}} dx\, F(x, s) \int_0^{xs} ds_+ \int_0^{(\sqrt{xs}-\sqrt{s_+})^2} ds_- \rho_W(s_+)\rho_W(s_-)\sigma_0^{\mathrm{CC03}}(s; s+, s-) \,. \tag{1.100}$$

Figure 1.9 compares the lowest order calculations using GENTLE [47]. One observes that both the finite width and ISR effects lead to a broadening of the production threshold.

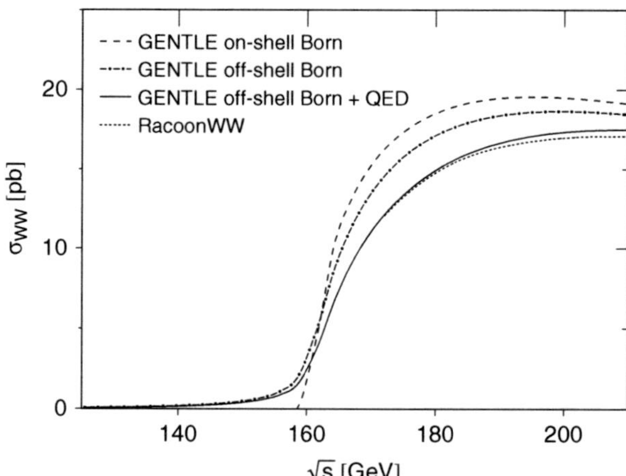

Fig. 1.9 Cross-section of W-pair production in e^+e^- collisions at lowest order, including W width effects, with ISR corrections calculated with GENTLE and the full $O(\alpha)$ calculation with RacoonWW

To eventually match the precision of the LEP measurements more complete radiative corrections need to be taken into account. The most recent predictions available in the RacoonWW [48] and KandY [49] Monte Carlo programs contain electroweak corrections at $O(\alpha)$. This brings the theoretical uncertainties on the cross-section to the level of 0.5% at centre-of-mass energies above 180 GeV. Around the threshold, $\sqrt{s} \approx 2M_W$, the precision is only in the order of 2%. The improvement in accuracy in the higher energy range is due to the so-called leading or double-pole approximations (LPA/DPA) [50] which are applied to treat the virtual radiative corrections. In these approximations the matrix element is expanded around the resonant poles in powers of $\Gamma_W/(M_W\beta)$. The expansion is therefore only valid when the velocity β is sufficiently large, i.e. well above the W-pair threshold.

Recent results with better accuracy for the threshold region are obtained in an effective field theoretical approach [51]. In the analysis of LEP data they are not yet applied. But they may become important when a future e^+e^- linear collider will be operated at the W-pair threshold and large statistics data samples are collected.

1.7 Z and W Boson Production at Tevatron and the LHC

At $p\bar{p}$ and pp colliders the electroweak gauge bosons are produced as a single particle or in pairs through a parton-parton process, for example the Drell-Yann production of W and Z bosons, $q\bar{q} \to Z$ and $\overline{qq'} \to W$. The total cross-section of a certain process $pp \to X$ at a centre of mass energy $\sqrt{s}$ can be written as:

$$\sigma = \sum_{i,j} \int \hat{\sigma}_{ij}(\hat{s}, \mu_f, \mu_r) \int_0^1 \int_0^1 f_i\left(x_1, \mu_f^2\right) f_j\left(x_2, \mu_f^2\right) \delta(\hat{s} - x_1 x_2 s)\, dx_1\, dx_2\, d\hat{s}\ . \tag{1.101}$$

The different quantities in the equation are

$$\begin{aligned}
\hat{\sigma}_{ij} &= \text{parton-parton cross-section } i + j \to X \\
\sqrt{\hat{s}} &= \text{reduced centre-of-mass energy of the parton reaction } i + j \to X \\
x_1 &= \text{energy fraction of parton } i;\ x_1 = \frac{2E_1}{\sqrt{s}} \\
x_2 &= \text{energy fraction of parton } j;\ x_2 = \frac{2E_2}{\sqrt{s}} \\
f_i(x) &= \text{parton distribution function (PDF) for parton } i \text{ (same for } j\text{)} \\
&= \text{probability to find parton } i \text{ with energy fraction } x \text{ inside the proton} \\
\mu_f &= \text{factorisation scale} \\
\mu_r &= \text{renormalisation scale}
\end{aligned}$$

The partons i and j may be quarks $(u, d, s, c, b, t, \bar{u}, \bar{d}, \bar{s}, \bar{c}, \bar{b}, \bar{t})$ and gluons inside the two protons of the colliding beams. The delta function $\delta(\hat{s} - x_1 x_2 s)$ ensures the energy conservation. In the very simplified case of the production of a narrow resonance the parton cross-section can be written as

$$\hat{\sigma}_{ij}(\hat{s}) = \sigma_{ij}\delta(\hat{s} - M^2)M^2 \tag{1.102}$$

where M is the mass of the resonance and σ_{ij} the (constant) cross-section of the reaction $i + j \rightarrow X$ at the peak of the resonance. The total cross-section is now simplified to

$$\sigma = \sum_{i,j} \sigma_{ij} M^2 \times L_{ij} \tag{1.103}$$

with the parton-parton luminosity

$$L_{ij} = \frac{1}{s} \int_{\frac{M^2}{s}}^{1} \frac{1}{x} f_i(x) f_j\left(\frac{M^2}{xs}\right) dx \tag{1.104}$$

neglecting dependencies on factorisation and renormalisation scales in this notation. The rapidity of the resonance X is defined as

$$y = \frac{1}{2} \log \frac{E - p_L}{E + p_L} \tag{1.105}$$

with energy E and longitudinal momentum p_L. It is well approximated by the more commonly used pseudo-rapidity

$$\eta = \frac{1}{2} \log \frac{|\mathbf{p}| - p_L}{|\mathbf{p}| + p_L} = -\log \tan \frac{\theta}{2}, \tag{1.106}$$

which can be directly measured in terms of the polar angle θ. The energy fraction, x, of the partons that produce the massive decay product at rapidity, y, is given by

$$x = \frac{M}{\sqrt{s}} e^{\pm y} \tag{1.107}$$

The differential rapidity distribution of X is therefore to lowest order proportional to the product of the parton density functions:

$$\frac{d\sigma}{dy}(pp \rightarrow X) = \sum_{i,j} \sigma_{ij} \frac{M^2}{s} f_i\left(\frac{M}{\sqrt{s}} e^{y}\right) f_j\left(\frac{M}{\sqrt{s}} e^{-y}\right) \tag{1.108}$$

Figure 1.10 shows the result of a proper calculation for W^+ and W^- production at the LHC simulated with a leading-order Monte Carlo program. One can observe a significant difference between the two charged bosons due to different quark contributions to the PDFs. This difference is planned be used in the determination of the parton luminosity at the LHC [52].

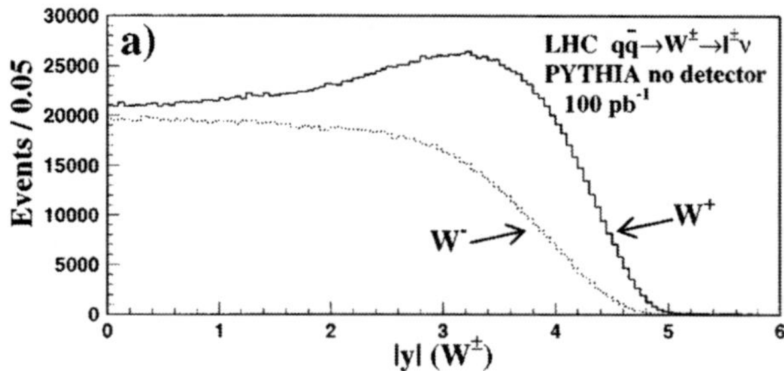

Fig. 1.10 Rapidity distribution at Monte Carlo generator level for W^+ and W^- production at LHC energies, taken from [52]

The simplified picture must be extended by taking W and Z width effects into account, as well as QCD and QED radiative corrections. Taking higher order corrections at next-to-leading order (NLO) and beyond into account, typically reduces the scale dependencies on μ_r and μ_f of the cross-section predictions (see e.g. Sect. 6.1 and 6.3). The production of W and Z bosons accompanied by jets with large transverse momentum is also an important source of background for searches for new particles at hadron-hadron colliders.

Production cross-sections for various Standard Model processes are shown in Fig. 1.11, including some examples for vector bosons with exclusive jet production [53]. Exclusive W/Z+jets cross-sections for up to three jets are recently available at NLO precision [54, 59]. Higher order Monte Carlo generators apply a matching of the fixed order QCD calculations to traditional parton shower models [55–58]. Thus, measurements of W/Z+jet final states at the Tevatron are reasonably well described by the theoretical predictions [53, 59]. This gives confidence that predictions for LHC energies can also be trusted at the percent level.

1.8 Standard Model Higgs Boson Production and Decay at the LHC

At the proton-proton collider LHC the Higgs boson is produced in several processes [10]:

- gluon fusion: $gg \rightarrow H$
- vector boson fusion: $qq \rightarrow qq + W^*W^*, Z^*Z^* \rightarrow qq + H$
- Higgs-strahlung off W or Z: $qq \rightarrow W, Z \rightarrow W, Z + H$
- Higgs bremsstrahlung off a top quark: $qq, gg \rightarrow t\bar{t} + H$

The corresponding Feynman diagrams are sketched in Fig. 1.12.

Gluon fusion is the by far dominating process with the highest cross-section over the whole Higgs mass range. Although the massless gluons do not couple directly to the Higgs, production via triangular quark loops is well possible. The large quark-mass coupling compensates the dynamic suppression due to the loop diagram. To

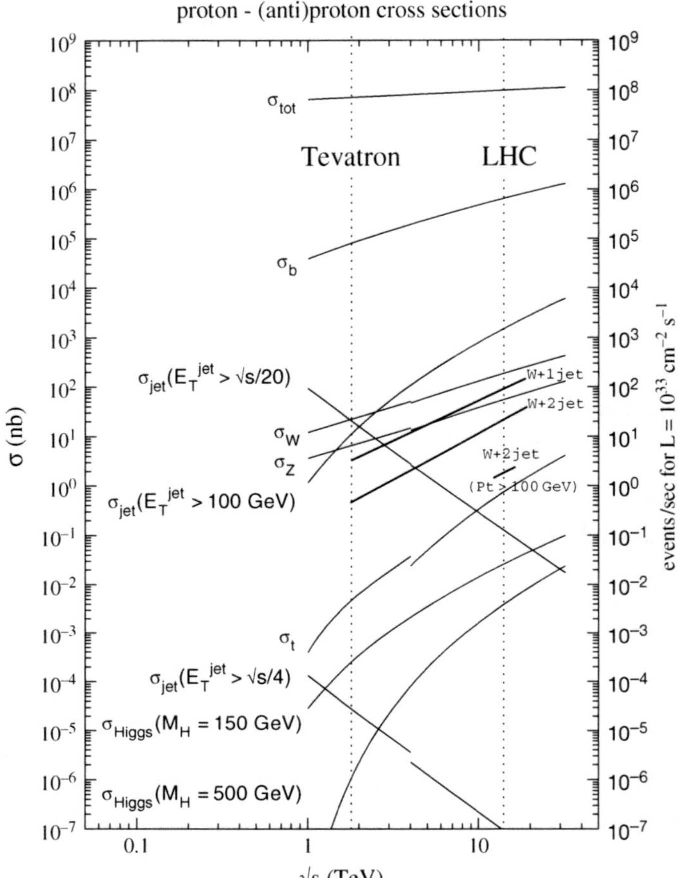

Fig. 1.11 Standard Model cross-sections at Tevatron and LHC energies, calculated at NLO precision. The discontinuities are due to the differences in parton content between $p\bar{p}$ and pp collisions. The *lines* indicated for W + 1,2 jets correspond to jets with $p_T > 20$ GeV and $|\eta| < 2.5$ [53]

lowest order the partonic $gg \to H$ cross-section can be written as [60]

$$\hat{\sigma}_{LO}(gg \to H) = \frac{\pi^2}{8M_{\rm H}} \Gamma_{LO}(H \to gg) \delta\left(\hat{s} - M_{\rm H}^2\right) \tag{1.109}$$

with

$$\Gamma_{LO}(H \to gg) = \frac{G_{\rm F} \alpha_s^2 M_{\rm H}^3}{36\sqrt{2}\pi^3} \left| \frac{3}{4} \sum_q A_q^H(\tau_q) \right|^2 . \tag{1.110}$$

The zero width approximation may be improved by substituting

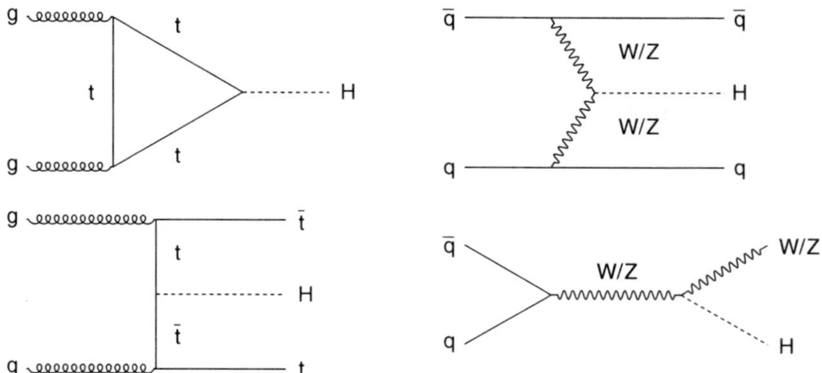

Fig. 1.12 Feynman diagrams for Higgs production showing gluon-gluon fusion and vector-boson fusion processes, as well as the associated Higgs production with top quarks and weak vector bosons

$$\delta\left(\hat{s} - M_{\mathrm{H}}^2\right) \rightarrow \frac{1}{\pi} \frac{\hat{s}\Gamma_{\mathrm{H}}/M_{\mathrm{H}}}{\left(\hat{s} - M_{\mathrm{H}}^2\right)^2 + (\hat{s}\Gamma_{\mathrm{H}}/M_{\mathrm{H}})^2} \tag{1.111}$$

The gluonic Higgs width $\Gamma_{LO}(H \rightarrow gg)$ is expressed in terms of form factors that depend on the squared Higgs-to-quark mass ratio $\tau_q = M_{\mathrm{H}}^2/4m_q^2$ [10]:

$$A_q^H(\tau_q) = \frac{2}{\tau_q^2}[\tau_q + (\tau_q - 1)f(\tau_q)] \tag{1.112}$$

$$f(\tau) = \begin{cases} \arcsin^2 \sqrt{\tau} & \tau \leq 1 \\ -\frac{1}{4}\left[\log \frac{1+\sqrt{1-\tau^{-1}}}{1-\sqrt{1-\tau^{-1}}} - i\pi\right]^2 & \tau > 1 \end{cases} \tag{1.113}$$

For small quark masses the form factor vanishes, and it approaches a value of $\frac{3}{4}$ for $m_q \gg M_{\mathrm{H}}$. This formula is also valid in extensions of the Standard Model, where higher mass fermions may appear in the loop and further increase the production cross-section. In the Standard Model, the main contribution is from the top quark loop. For Higgs masses below $2m_{\mathrm{t}}$ the infinite top mass approximation $m_{\mathrm{t}} \rightarrow \infty$ agrees with the full result within 10%, as can be seen in Fig. 1.13.

Corrections to the leading order cross-section are necessary because higher order QCD processes generally change the lowest order results significantly. For total cross-section calculations the corrections are usually phrased in terms of a K factor, which takes NLO or even higher order effects into account, e.g. $K = \sigma_{\mathrm{NLO}}/\sigma_{\mathrm{LO}}$ in case of NLO corrections. The LO cross-section can be written in the following form

$$\sigma(gg \rightarrow H) = \sigma_0^H \tau_H \frac{d\mathcal{L}^{gg}}{d\tau_H} \tag{1.114}$$

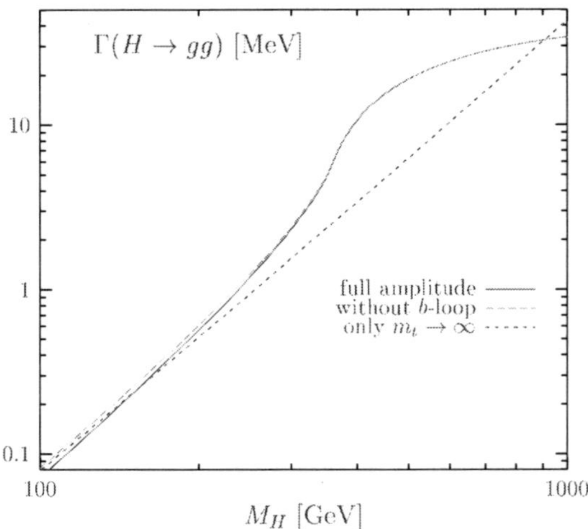

Fig. 1.13 Gluonic width of the Higgs boson including all quarks in the triangular loop, excluding the b-quark, and in the large m_t limit [10]

where $d\mathcal{L}^{gg}/d\tau_H$ denotes the gg luminosity of the pp collider as function of the Drell-Yann variable $\tau_H = M_H^2/s$, where s is the invariant collider energy squared. This notation is helpful when adding NLO terms, which arise through the real and virtual contributions, as shown in Fig. 1.14. The NLO cross-section is then given by:

$$\sigma(gg \to H) = \sigma_0^H \tau_H \left[1 + C^H \frac{\alpha_s}{\pi}\right] \frac{d\mathcal{L}^{gg}}{d\tau_H} + \Delta\sigma_{gg}^H + \Delta\sigma_{gq}^H + \Delta\sigma_{q\bar{q}}^H \tag{1.115}$$

The coefficient C^H is the finite part of the virtual two-loop corrections [10], which are known to order α_s^5. The $\Delta\sigma$ terms are the hard contributions from gluon radiation in gg and gq scattering and $q\bar{q}$ annihilation. The corresponding K-factors for these terms are shown in Fig. 1.14. The virtual K_{virt} and the K_{gg} factors are the largest and in the order of 50%, while the others do not contribute much. The total correction at LHC energies is between 60 and 90% for low and high Higgs mass ranges, respectively.

Recent calculations even include NNLO, soft N^{3}LO and N^3 leading-log (N^{3}LL) calculations [61]. It turns out that NNLO corrections are still relatively large. Only at the following order the perturbation series starts to converge and yields smaller contributions (see Fig. 1.14). At all orders the dependence on the Higgs mass is large, and the gluon fusion cross-section at the LHC drops from about 60 pb for low M_H to below 10 pb for large M_H.

For the description of the transverse momentum, p_T, and rapidity, η, dependence of the $gg \to H$ process at higher orders Monte Carlo techniques are used. Furthermore, this allows the application of more realistic phase space cuts that are close to event selections applied on detector level. Calculations with the NNLO program FEHiP [62] have shown that NNLO effects can change the cross-section

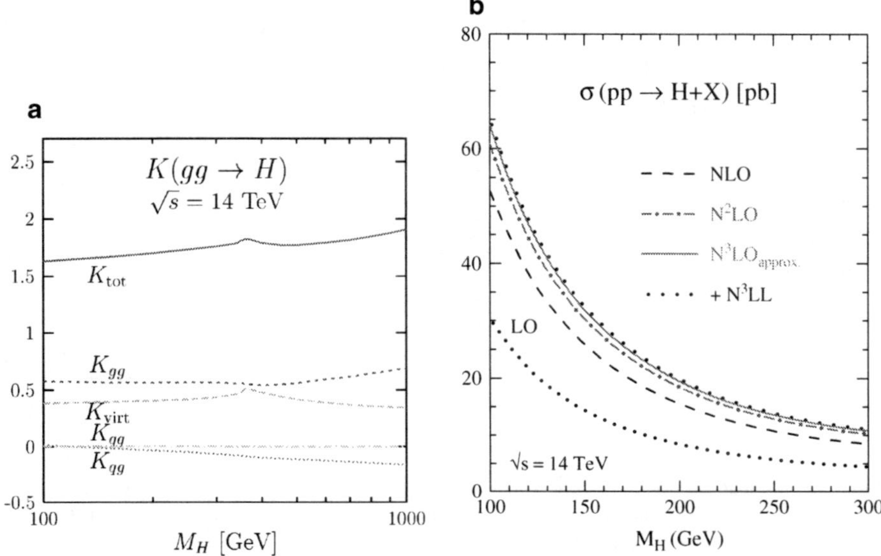

Fig. 1.14 (**a**) K-factors for the $gg \to H$ production cross-section at NLO [10]. (**b**) N^{3}LO calculations of the $gg \to H$ production cross-section [61]

within cuts by up to 5% [63]. Standard Monte Carlo programs like Pythia [64] or MC@NLO [58], only include LO or NLO effects, but may be improved to NNLO level by event reweighting [65].

Vector boson fusion (VBF) has a lower total cross-section than the gluon fusion process. But it provides the additional signature of quark jets with small transverse momentum, p_T, that can be identified by the LHC detectors. In the longitudinal vector boson approximation [66] the total partonic cross-section is calculated to be [10]:

$$\hat{\sigma}_{LO}(qq \to qqH) = \frac{G_F^3 M_V^4 N_c}{4\sqrt{2}\pi^3} C_V \left\{ \left(1 + \frac{M_H^2}{\hat{s}}\right) \log \frac{\hat{s}}{M_H^2} - 2 + 2\frac{M_H^2}{\hat{s}} \right\} , \quad (1.116)$$

where $N_c = 3$ denotes the colour factor and M_V the vector boson mass. The factor C_V contains the quark-boson coupling constants:

$$C_Z = \left(\left(g_V^{q1}\right)^2 + \left(g_A^{q1}\right)^2 \right) \left(\left(g_V^{q2}\right)^2 + \left(g_A^{q2}\right)^2 \right) , \quad C_W = 1 . \quad (1.117)$$

Because of the larger charged couplings the WW fusion is one order of magnitude larger than the ZZ fusion in this approximation. More complete calculations include all polarisations of the intermediate bosons, like the one displayed in Fig. 1.15, and NLO Monte Carlo programs are available [67].

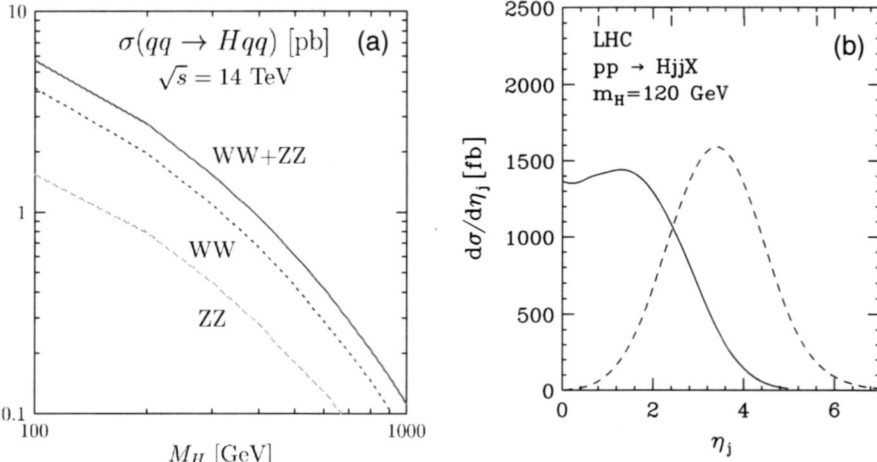

Fig. 1.15 (**a**) Vector boson fusion cross-section $\sigma(qq \to V^*V^* \to Hqq)$ at leading order [10]. (**b**) Pseudo-rapidity distributions of the quark jets in the VBF process. The most central jet is shown as a *solid line*, the most forward one as a *dashed line* [68]

Especially interesting is the kinematic behaviour of the quarks and bosons in the VBF process. The direction of the bosons emitted from the initial partons is in general close to the actual parton direction and their energies are of the order of the Higgs mass. Therefore the remaining parton quarks keep practically all their initial energy of about 1 TeV (at LHC) and have small transverse momentum, p_T. This also means that the hadronic quark jets are produced in the forward region. When expressed with the pseudo-rapidity, $\eta = -\log\tan\frac{\theta}{2}$, which is a function of the polar angle θ with respect to the colliding particles, values in the range $1 < \eta < 5$ are preferred, as illustrated in Fig. 1.15. Since there is no colour flow between the two initial parton quarks a so-called rapidity gap is expected to be observed in VBF production, which means that the hadronic activity in the central η range is reduced. This feature is used to reduce background, mainly from $t\bar{t}$ events which are produced more centrally.

In the low Higgs mass region, $M_{\mathrm{H}} < 150$ GeV, also the associated production with W and Z bosons, $q\bar{q} \to WH, ZH$ has a sizable production rate. To lowest order the partonic cross-section is given by:

$$\hat{\sigma}(q\bar{q} \to VH) = \frac{G_{\mathrm{F}}^2 M_V^4}{72\pi\hat{s}} C_V \lambda^{1/2}\left(M_V^2, M_{\mathrm{H}}^2, \hat{s}\right) \frac{\lambda\left(M_V^2, M_{\mathrm{H}}^2, \hat{s}\right) + 12M_V^2/\hat{s}}{\left(1 - M_V^2/\hat{s}\right)^2} \tag{1.118}$$

with the coupling factors for $V = W, Z$:

$$C_Z = \left(g_{\mathrm{V}}^q\right)^2 + \left(g_{\mathrm{A}}^q\right)^2 \quad , \; C_W = 1 \tag{1.119}$$

and the two-body phase space function

$$\lambda(x, z, y) = \left(1 - \frac{x}{z} - \frac{y}{z}\right)^2 - 4\frac{xy}{z^2} . \qquad (1.120)$$

Since it is a two-body decay the Higgs and the vector boson are produced back-to-back in the $q\bar{q}$ rest frame, which may be used in the search for the Higgs boson. The dependence of $\sigma(q\bar{q} \to VH)$ on the Higgs mass at the LHC is shown in Fig. 1.16, including NNLO QCD and electroweak radiative corrections.

The associated Higgs production with heavy quarks is like the associated vector boson production mainly important for low Higgs masses. There are ten leading order Feynman diagrams, since not only quark annihilation $q_1\bar{q}_1 \to g \to q_2\bar{q}_2H$ contributes but also gluon fusion $gg \to q_2\bar{q}_2H$ with s- and t-channel graphs. A closed expression is therefore quite involved. More details can be found in [10]. The evolution of $t\bar{t}H$ production with M_H for LO and NLO calculations is given in Fig. 1.16.

Figure 1.17 shows a summary of the different cross-sections for the various Higgs production mechanisms. At the LHC, gluon fusion clearly dominates and VBF is very important in all M_H ranges.

The decay of the Higgs boson eventually determines the search strategy at the LHC. The gluonic decay width already played a role in the gluon fusion process. However, due to the multi-hadronic environment caused by the underlying event and pile-up events, purely hadronic Higgs decays are very difficult to detect.

The other loop induced decay into two real photons, $H \to \gamma\gamma$, is more important because it has a clear detector signature. The decay width at lowest order is similar

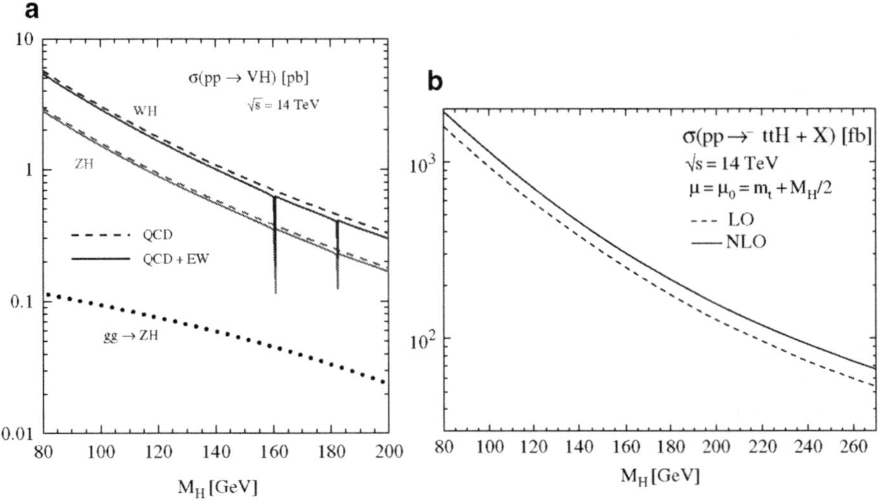

Fig. 1.16 (**a**) Higgs cross-sections for associated boson production. The *solid line* includes both NNLO QCD and electroweak radiative corrections, the *dashed line* only includes NNLO QCD effects. At this order of the perturbation series, also $gg \to ZH$ production contributes, which is indicated separately [69]. (**b**) Predictions of the $t\bar{t}H$ production at LHC energies at LO and NLO [70]

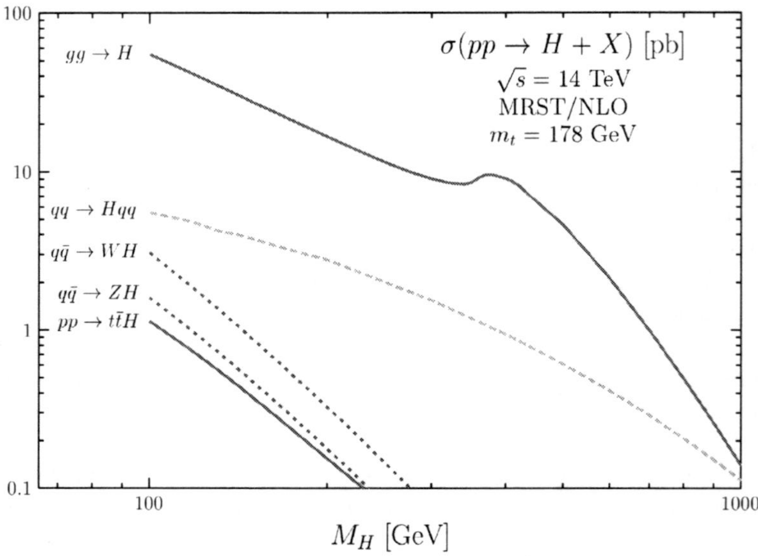

Fig. 1.17 Higgs production cross-sections at NLO [10]

to Eq. (1.110), except that the coupling to the final state is electromagnetic and all charged particles are included in the loop:

$$\Gamma_{LO}(H \to \gamma\gamma) = \frac{G_F \alpha_{QED}^2 M_H^3}{128\sqrt{2}\pi^3} \left| \sum_f N_c Q_f A_f^H(\tau_f) + A_W^H(\tau_W) \right|^2 \quad (1.121)$$

The additional term

$$A_W^H(\tau) = -\frac{1}{\tau^2}[2\tau^2 + 3\tau + 3(2\tau - 1)f(\tau)] \quad (1.122)$$

is from W bosons in the loop. The fermionic amplitude $A_f^H(\tau_f)$ and the functions $f(\tau)$ are defined in Eq. (1.112). The colour factor, N_c, equals 3 for quarks and 1 for leptons.

The photonic decay width is much smaller than the gluonic one, as shown in Fig. 1.18. Both decrease fast with increasing Higgs mass when the decay channels to the heavy vector bosons open. This is also the case for the decay into low mass fermions.

The fermionic Higgs decay width at lowest order is given by

$$\Gamma(H \to f\bar{f}) = \frac{G_F N_c}{4\sqrt{2}\pi} M_H m_f^2 \beta_f^3 \quad (1.123)$$

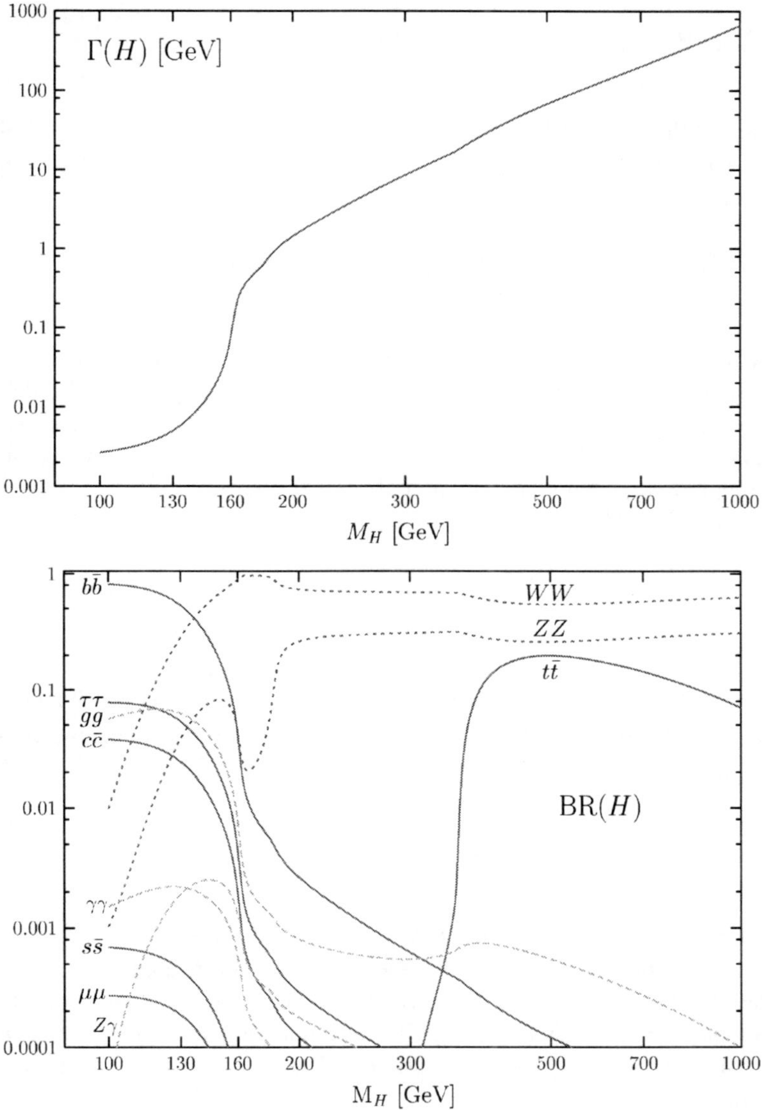

Fig. 1.18 Higgs total decay width (*top*) and branchings fractions (*bottom*). Two-loop QCD and leading electroweak corrections are included [10]

with the velocity of the fermions, $\beta_f = \sqrt{1 - 4m_{\mathrm{f}}^2/M_{\mathrm{H}}^2}$. Since the width is proportional to the fermion mass, mainly b and t quarks need to be considered. In this case, also QCD corrections need to be taken into account. If the Higgs mass is much larger than the quark mass, $M_{\mathrm{H}} \gg m_q$, which is the case for the b quark, one can approximate at NLO:

$$\Gamma_{\mathrm{NLO}}(H \to q\bar{q}) \approx \frac{3G_{\mathrm{F}}}{4\sqrt{2}\pi} M_{\mathrm{H}} m_q^2 \left\{ 1 + \frac{4}{3}\frac{\alpha_s}{\pi} \left(\frac{9}{4} + \frac{3}{2} \log \frac{m_q^2}{M_{\mathrm{H}}^2} \right) \right\} \tag{1.124}$$

Also for the top quark, where the previous approximation does not hold any more, QCD corrections are formally added by a Δ_H^t term that depends on the top velocity β_t:

$$\Gamma_{\mathrm{NLO}}(H \to t\bar{t}) = \frac{3G_{\mathrm{F}}}{4\sqrt{2}\pi} M_{\mathrm{H}} m_{\mathrm{t}}^2 \beta_t^3 \left\{ 1 + \frac{4}{3}\frac{\alpha_s}{\pi} \Delta_H^t(\beta_t) \right\} \tag{1.125}$$

Due to the kinematics the branching fractions for low mass Higgs bosons, $M_{\mathrm{H}} < 150$ GeV, is dominated by the $b\bar{b}$ decay, as can be seen in Fig. 1.18. Also the decay to tau pairs is important. Top pairs are clearly only produced beyond the $2m_{\mathrm{t}}$ mass threshold. However, Higgs decays into vector boson pairs are still dominating the branching ratio in the high mass region $M_{\mathrm{H}} > 150$ GeV.

The Higgs partial decay width into two real vector bosons is given by

$$\Gamma(H \to VV) = \frac{G_{\mathrm{F}} M_{\mathrm{H}}^3}{16\sqrt{2}\pi} \delta_V \sqrt{1-4x}(1-4x+12x^2) \tag{1.126}$$

with $\delta_W = 2$, $\delta_Z = 1$ and the mass ratio $x = \frac{M_V^2}{M_{\mathrm{H}}^2}$. This means that for very large Higgs masses the decay width into W bosons is two times larger than the one into Z bosons. Also interesting is the longitudinal polarisation fraction

$$\frac{\Gamma_L}{\Gamma_T + \Gamma_L} = \frac{1-4x+4x^2}{1-4x+12x^2} \tag{1.127}$$

which approaches 1 for large $M_{\mathrm{H}} \gg M_V$. The W and Z boson are therefore practically 100% longitudinally polarised in the high Higgs mass range.

When the Higgs mass is not large enough to decay into on-shell vector bosons, off-shell production needs to be taken into account, with the subsequent decay into lepton pairs, $H \to VV^* \to$ ffff. Details of this decay mode are described in Chap. 7.

The total Higgs decay width as a function of M_{H} is displayed in Fig. 1.18, together with the branching fractions into the various final states.

1.9 Comparison of Theory and Experiment

Theories need to be compared with physics data to either approve or falsify their predictions. An important technique is the Monte Carlo modelling of physics processes, the subsequent simulation of their signatures in the experimental setup and the final comparison to measured data distributions and interaction rates. Many Monte Carlo generators are available and each has its specific advantages and physics domains where they are applied.

Pythia [64] and Herwig [71] are multi-purpose generator and are used to simulate e^+e^-, ep, and hadron collision events. In collisions involving protons, the distribution of the partons inside the proton are taken from internal or external PDF libraries, like LHADPDF [72]. The most common PDF sets are CTEQ6 [73] and MRST [74] which are to be matched to the proper order of the QCD perturbation series of the Monte Carlo process. Pythia and Herwig (in combination with JIMMY [75]) also include the simulation of beam-remnants, underlying event and multiple interactions for pp and $p\bar{p}$ processes. Both programs are used for simulating the fragmentation and hadronisation of quarks and gluons and can be run together with other event generators.

At LEP, four-fermion final states are mostly generated with the programs KandY [49], RacoonWW [48], GRC4f [76], EXCALIBUR [77], YFSZZ [78], and WPHACT [79]. Fermion-pair production is simulated with KORALZ [80] and its successor KK2f [81] and Bhabha scattering with BHWIDE [82] and TEEGG [83]. Hadronic 2-photon events are best described by PHOJET [84] and TWOGAM [85], while leptonic events of the same kind are usually generated with DIAG36 [86] and LEP4F [87]. Photon radiation off leptons can be produced with PHOTOS [88] and tau lepton decays are modelled in TAUOLA [89].

For proton-proton collisions at the LHC, the Sherpa [56] and Alpgen [57] programs implement parton-shower matching and are able to describe W/Z production with up to four and five jets, as well as VBF and b-quark associated Higgs production. Alpgen is applied in photon pair production, like the multi-purpose tool MadGraph/MadEvent [90], which also describes vector boson and vector boson pair production. The purpose of the AcerMC [91] package is the generation of $Zb\bar{b}/Zt\bar{t}$ as well as top pair events.

The MC@NLO [58] event generator is one of the few Monte Carlo tools including full NLO corrections to a selected set of processes in a consistent way, like inclusive W or Z production, $t\bar{t}$ production, electroweak boson pair production, as well as Higgs boson production and decay to W^+W^- and $\gamma\gamma$ final states.

Production cross-sections of electroweak bosons at NNLO accuracy are available using FEWZ [92]. NLO calculations of the production of W and Z bosons and two jets with or without heavy quark tag can be performed with the MCFM [93] program and higher order corrections to $\gamma\gamma$ production are commonly done with RESBOS [94].

1.9.1 Hadronisation Models

There are three main Monte Carlo programs that are used for modelling fragmentation and hadronisation of quarks and gluons: Pythia [64], Herwig [71], and Ariadne [95]. The quarks and gluons are usually the result of the preceding Monte Carlo generation step.

In Pythia, the leading order hard scattering process is completed by the parton shower formalism to incorporate QED and QCD radiative leading-log corrections. In both initial- and final-state, the showers develop according to the branchings $e \to e\gamma$, $q \to qg$, $q \to q\gamma$, $g \to gg$, and $g \to q\bar{q}$. The rates are proportional to the integral $\int P_{a\to bc}(z)\ , dz$, with the splitting kernel $P_{a\to bc}(z)$ which depends on the energy fraction $z = E_b/E_a$ carried by the splitting product b. The other particle has energy $(1-z)$ after the splitting process. In the final state, Pythia is evolving the showers from a virtuality scale scale Q^2_{max} down to a lower scale Q^2_0. Time ordering is done either according to the mass, m, of the shower partons or in recent Pythia versions according to p_T, where $p_T^2 = z(1-z)m^2$.

The hadronisation package of Pythia can be used independently from the main generator. It applies the string fragmentation model to the parton shower products or to inputs from external generators. The string picture starts from the assumption that the energy between a colour dipole, like a $q\bar{q}$ pair for example, is linearly increasing with the distance between the charges. When the charges move further apart the energy stored in the string increases and it eventually breaks to form a new colour charge pair. A $q\bar{q}$ pair may split into two colour singlets $q\bar{q}'$ and $q'\bar{q}$. This process continues until only on-mass-shell hadrons remain. The subsequent decays of the hadrons are also treated by Pythia.

Several parameters are available to adjust the model to measured data. Especially the tuning of the parton shower and fragmentation are important, and in case of hadron collisions also the underlying event structure. For the parton showers the Lund fragmentation function is typically used for light, uds, flavours and the Petersson function for heavy, c and b, flavours with their tuned parameters. Furthermore, the Λ_{QCD} value used for the running of the strong coupling α_s in parton showers can be adjusted, as well as the parton shower cut-off value Q_0. In general, each experiment is individually trying to obtain the best description of the hadronic data distributions.

For LHC studies, the recent implementation of parton showering, commonly known as p_T-ordered showering, is used together with the new underlying event model where the phase-space is interleaved/shared between initial-state radiation (ISR) and the underlying event.

The Herwig program also treats quark fragmentation according to the parton shower model, fragmentation is however performed differently. In the Herwig cluster fragmentation model quarks and gluons from the parton showers combine locally into clusters. They are much less extended and less massive objects than strings. Only singlet combinations of partons are allowed to form clusters. These decay quasi isotropically into a small number of hadrons each. Like for Pythia, measured hadronic distributions are used to tune the Herwig model parameters, for example the QCD scale Λ_{QCD} or the cluster mass parameters which describe the cluster fission.

The Ariadne program applies the dipole-cascade model to the fragmentation of quarks and gluons. The emission of a gluon g_1 from a quark anti-quark pair $q\bar{q}$ is modelled as radiation from a colour dipole between the q and the $\bar{q}$. The two new dipoles qg_1 and $g_1\bar{q}$ again radiate softer gluons. Radiation from the $q\bar{q}$ dipole is suppressed by the colour factor $1/N_c^2$. The strong coupling α_s in the differential cross-sections $d\sigma/(dx\,dx_g)$ for the processes $q\bar{q} \to q\bar{q} + g$, $qg \to qg + g$, and $gg \to gg + g$ is evaluated at the transverse momentum scale, p_T^2, of the emission. The ordering of the gluon emission is also arranged according to the p_T^2 scale, where Sudakov form factors describe the probability of having emissions at a higher scale. Gluon splitting $g \to q\bar{q}$ is also possible and competes with the emission of another gluon. QED photon radiation in the cascade are treated similarly to the QCD gluon emission, however the photon emission is less probable since the electromagnetic coupling constant is much smaller than the strong coupling. Eventually, at the end of the cascade, the Pythia hadronisation model is applied to form the final state hadrons. The main parameters that can be tuned in Ariadne to describe hadronic data are, for example, Λ_{QCD} and a p_t cut-off parameter.

1.9.2 Detector Simulation

The simulation of the interaction of the final state particles with the detector is important to deduct the properties of the underlying physics processes from the measurement. The experiments at LEP and the Tevatron used the GEANT3 [96] FORTRAN program to calculate the particle trajectories and their energy depositions in the detector. Effects like ionisation energy loss, multiple scattering, electromagnetic showering and hadronic interaction with matter are implemented. The LHC experiments moved to the C++ coded GEANT4 [97] software, which is an evolution of GEANT3. The detailed detector simulations are in general very computing time consuming. Faster parameterisations [98] of the detector behaviour are therefore used in some applications under the condition that their precision is sufficient for the physics measurement.

References

1. S. L. Glashow, Nucl. Phys. **22** (1961) 579; S. Weinberg, Phys. Rev. Let. **19** (1967) 1264; A. Salam, *Elementary Particle Theory*, (ed) N. Svartholm, Stockholm, *Almquist and Wiksell*, 1968, p. 367; G. t'Hooft, Nucl. Phys. **B 35** (1971) 167; G. t'Hooft and M. Veltman, Nucl. Phys. **B 44** (1972) 189.
2. H. Fritzsch, M. Gell-Mann and H. Leutwyler, Phys. Lett. **B 47** (1973) 365; H. D. Politzer, Phys. Rev. Let. **30** (1973) 1346; D. Gross and F. Wilczek, Phys. Rev. Let. **30** (1973) 1343; S. Weinberg, Phys. Rev. Let. **31** (1973) 494.
3. The Particle Data Group, C. Amsler, et al., Phys. Lett. **B 667** (2008) 1.
4. M. Kobayashi and T. Maskawa, Prog. Th. Phys. **49** 2 (1973) 652.
5. P. W. Higgs, Phys. Lett. **12** (1964) 132; *idem*, Phys. Rev. Lett. **13** (1964) 508; *idem*, Phys. Rev. **145** (1966) 1156; F. Englert and R. Brout, Phys. Rev. Lett. **13** (1964) 321; G. S. Guralnik, C. R. Hagne and T. W. B. Kibble, Phys. Rev. Lett **13** (1964) 585.
6. See for example: U. Dore and D. Orestano, Rep. Prog. Phys. **71** (2008) 106201; T. M. Raufer, *Neutrino Oscillation Results from MINOS and MiniBooNE*, arXiv:0808.0392v1; T. Schwetz, M. Tortola and J. W. F. Valle, New. J. Phys. **10** (2008) 113011, arXiv:0808.2016v2.

7. Z. Maki, M. Nakagawa and S. Sakata, Prog. Theor. Phys. **28** (1962) 870; B. Pontecorvo, Zh. Eksp. Teor. Fiz. **53** (1967) 1717. See [3] for a summary.
8. H. V. Klapdor-Kleingrothaus, et al., Mod. Phys. Lett. **A 16** (2001) 2409; C. E. Aalseth, et al., Mod. Phys. Lett. **A 17** (2002) 1475; H. V. Klapdor-Kleingrothaus and I. V. Krivosheina, Phys. Lett. **A 21** (2006) 1547; The latter reference claims evidence for a signal, which is however not fully accepted in the community and needs confirmation by future experiments. A present status is e.g. reviewed in: A.S. Barabash, *Double beta decay: present status*,arXiv:0807.2948v1.
9. M. Chanowitz, *Strong WW scattering at the end of the 90's: theory and experimental prospects*, hep-ph/9812215v1.
10. For a review, see for example: Djouadi, A., *The Anatomy of Electro-Weak Symmetry Breaking. I: The Higgs boson in the Standard Model*, Phys. Rept. **457** (2008) 1, hep-ph/0503172; Djouadi, A., *The Anatomy of Electro-Weak Symmetry Breaking. II: The Higgs bosons in the Minimal Supersymmetric Model*, Phys. Rept. 459 (2008) 1, hep-ph/0503173.
11. N. Cabibbo, L. Maiani, G. Parisi and R. Petronzio, Nucl. Phys. **B 158** (1979) 295; R. Dashen and H. Neuberger, Phys. Rev. Lett. **50** (1983) 1897; D. Callaway, Nucl. Phys. **B 233** (1984) 189; P. Hasenfratz and J. Nager, Z. Phys. **C 37** (1988) 477; J. Kuti, L. Lin, and Y. Shen, Phys. Rev. Lett. **61** (1988) 678; R. Chivukula and E. Simmons, Phys. Lett. **B 388** (1996) 788; U. M. Heller, M. Klomfass, H. Neuberger, P. Vranas, Nucl.Phys. **B 405** (1993) 555.
12. J. A. Casas, J. R. Espinosa and M. Quiros, Phys. Lett. **B 342** (1995) 171;Phys. Lett. **B 382** (1996) 374.
13. J. R. Espinosa, G. Giudice and A. Riotto, *Cosmological implications of the Higgs mass measurement*, arXiv:0710.2484v1.
14. T. Hambye and K. Riesselmann, Phys. Rev. **D 55** (1997) 7255; C. F. Kolda and H. Murayama, JHEP **0007** (2000) 035.
15. M. Veltman, Acta Phys. Pol. **B 12** (1981) 437.
16. see for example: A. Linde, *Inflationary Cosmology*, Lect. Notes Phys. **738** (2008) 1, arXiv:0705.0164v2.
17. F. Bezrukov and M. Shaposhnikov, *Standard Model Higgs boson mass from inflation: two loop analysis*, arXiv:0904.1537v.
18. S. Bethke, J.Phys. **G 26** R27, 2000; hep-ex/0004021.
19. CODATA Recommended Values of the Fundamental Physical Constants: P. J. Mohr, B. N. Taylor, and D. B. Newell, Rev. Mod. Phys. **80** (2008) 633.
20. M. Steinhauser, Phys. Lett **B 429** (1998) 158.
21. G. Montagna et al., Comp. Phys. Comm. **117** (1999) 278; A. B. Arbuzov et al., Comp. Phys. Comm. **174** (2006) 728.
22. H. Burkhardt and B. Pietrzyk, Phys. Rev. **D 72** (2005) 057501.
23. Tevatron Electroweak Working Group, for the CDF Collaboration, the DØ Collaboration, *Combination of CDF and D0 Results on the Mass of the Top Quark*, arXiv:0903.2503v1.
24. F. Jegerlehner, *Renormalizing the Standard Model*, Preprint, PSI-PR-91-08, 1991.
25. The ALEPH, DELPHI, L3, OPAL, SLD Collaborations, the LEP Electroweak Working Group, the SLD Electroweak and Heavy Flavour Groups, Phys. Rep. **427** (2006) 257.
26. The ALEPH, DELPHI, L3, OPAL, SLC Collaborations, the LEP Electroweak Working Group, the Tevatron Electroweak Working Group, and the SLD electroweak and heavy flavour groups, *Precision Electroweak Measurements and Constraints on the Standard Model*, CERN-PH-EP/2008-020; arXiv:0811.4682. Updates can be found at http://lepewwg.web.cern.ch/LEPEWWG/.
27. The ALEPH, DELPHI, L3, OPAL Collaborations and the LEP Higgs Working Group, Phys. Lett. **B 565** (2003) 61.
28. R. Barbieri, S. Ferrara and C. A. Savoy, Phys. Lett. **B 119** (1982) 343; A. H. Chamseddine, R. Arnowitt and P. Nath, Phys. Rev. Lett. **49** (1982) 970; L. Hall, J. Lykken and S. Weinberg, Phys. Rev. **D 27** (1983) 2359; P. Nath, R. Arnowitt and A. H .Chamseddine, Nucl. Phys. **B 227** (1983) 121; P. Nath, R. Arnowitt and A. H. Chamseddine, *Applied N=1 Supergravity*

Trieste lecture series, vol. I, World Scientific, 1984; H. P. Nilles, *Supersymmetry, Supergravity and Particle Physics*, Phys. Rep. **110** (1984) 1.
29. P. Fayet, Phys. Lett. **B 70** (1977) 461; Phys. Lett. **B 86** (1979) 272; Phys. Lett. **B 175** (1986) 471; M. Dine, A. E. Nelson, Y. Nir and Y. Shirman, Phys. Rev. **D 53** (1996) 2658; H. Baer, M. Brhlik, C. H. Chen and X. Tata, Phys. Rev. **D 55** (1997) 4463; H. Baer, P. G. Mercadante, X. Tata and Y. L. Wang, Phys. Rev. **D 60** (1999) 055001; S. Dimopoulos, S. Thomas and J. D. Wells, Nucl. Phys. **B 488** (1997) 39; J. R. Ellis, J. L. Lopez and D. V. Nanopoulos, Phys. Lett. **B 394** (1997) 354; and the review: G.F. Giudice, R. Rattazzi, *Theories with Gauge-Mediated Supersymmetry Breaking*, Phys. Rep. **322** (1999) 419, hep-ph/9801271v2.
30. The ATLAS Collaboration, G. Aad, et al., *Expected Performance of the ATLAS Experiment Detector, Trigger, Physics*, CERN-OPEN-2008-020, arXiv:0901.0512.
31. W. Beenakker, R. Hopker, M. Spira and P. M. Zerwas, Nucl. Phys. **B 492** (1997) 51; W. Beenakker, et al., Phys. Rev. Lett. **83** (1999) 3780; Prospino2, http://www.ph.ed.ac.uk/tplehn/prospino/.
32. The ATLAS Collaboration, *Detector and Physics Performance Technical Design Report*, CERN/LHCC/99-14/15, 1999.
33. B. C. Allanach, A. Djouadi, J. L. Kneur, W. Porod and P. Slavich, JHEP **0409** (2004) 044.
34. The CMS Collaboration, G. L. Bayatian, et al., CMS Physics Technical Design Report, Vol. II: Physics Performance, CERN/LHCC 2006-021.
35. see for example: M. Drees, R. Godbole and P. Roy, *Theory and Phenomenology of Sparticles*, World Scientific Publishing Company, 2004.
36. M. C. Chen, S. Dawson, C. B. Jackson, Phys. Rev. **D 78** (2008) 093001; A. Aranda, J. Hernandez-Sanchez, P. Q. Hung, *Implications of the discovery of a Higgs triplet on electroweak right-handed neutrinos*, arXiv:0809.2791v1; E. Ma, U. Sarkar, Phys. Rev. Lett. **80** (1998) 5716.
37. M. Gell-Mann, P. Ramond and R. Slansky, in Supergravity, (ed) by D. Freedman et al., North Holland (1979); T. Yanagida, Prog. Theor. Phys. **64** (1980) 1103.
38. M. Aoki and S. Kanemura, Phys. Rev. **D 77** (2008) 095009.
39. The DØ Collaboration, V. Abazov, et al., Phys. Rev. Lett. **101** (2008) 071803, and references therein.
40. M. Schmaltz and D. Tucker-Smith, Ann. Rev. Nucl. Part. Sci. **55** (2005) 229.
41. H. C. Cheng, *Little Higgs, Non-standard Higgs, No Higgs and All That*, arXiv:0710.3407v1.
42. see for example: C. T. Hill, E. H. Simmons, Phys. Rep. **381** (2003) 235; Erratum ibid. **390** (2004) 553.
43. C. Csaki, C. Grojean, L. Pilo and J. Terning, Phys. Rev. Lett. **92** (2004) 101802; C. Englert, B. Jager, M. Worek and D. Zeppenfeld, *Observing Strongly Interacting Vector Boson Systems at the CERN Large Hadron Collider*, arXiv:0810.4861v1.
44. A. Alboteanu, W. Kilian and J. Reuter, JHEP **0811** (2008) 010.
45. The ALEPH Collaboration, the DELPHI Collaboration, the L3 Collaboration, the OPAL Collaboration, the SLD Collaboration, the LEP Electroweak Working Group, the SLD electroweak, heavy flavour groups, Phys. Rept. **427** (2006) 257; hep-ex/0509008v3.
46. G. Altarelli,. T. Sjostrand and F. Zwirner, (eds), *Workshop on Physics at LEP2*, Yellow report CERN 96-01, 1996.
47. GENTLE version 2.0 is used. D. Bardin et al., Comp. Phys. Comm. **104** (1997) 161.
48. RacoonWW, A. Denner, S. Dittmaier, M. Roth and D. Wackeroth, Phys. Lett. **B 475** (2000) 127; A.Denner, S.Dittmaier, M. Roth and D.Wackeroth, *W-pair production at future e+e- colliders: precise predictions from RACOONWW*, hep-ph/9912447.
49. KandY runs concurrently KORALW version 1.51 and YFSWW3 version 1.16. S. Jadach et al., Comp. Phys. Comm. **140** (2001) 475.
50. see [48] and [49].
51. M. Beneke, et al., *Four-fermion production near the W pair production threshold*, CERN-PH-TH-07-107, arXiv:0707.0773v1 [hep-ph].
52. M. Dittmar, F. Pauss and D. Zürcher, Phys. Rev. **D 56** (1997) 7284.

53. J. M. Campbell, *Overview of the theory of W/Z + jets and heavy flavor*, arXiv:0808.3517v1, and references therein.
54. C. F. Berger, Z. Bern, L. J. Dixon, F. Febres Cordero, D. Forde, H. Ita, D. A. Kosower and D. Maitre, *One-Loop Multi-Parton Amplitudes with a Vector Boson for the LHC*, arXiv:0808.0941v1.
55. S. Catani, F. Krauss, R. Kuhn and B. R. Webber, JHEP **0111** (2001) 063.; M. L. Mangano, M. Moretti, F. Piccinini, R. Pittau and A. D. Polosa, JHEP **0307** (2003) 001; S. Frixione, P. Nason and B. R. Webber, JHEP **0308** (2003) 007.
56. T. Gleisberg, S. Hoeche, F. Krauss, A. Schaelicke, S. Schumann and J. Winter, JHEP **0402** (2004) 056.
57. M. L. Mangano, M. Moretti, F. Piccinini, R. Pittau and A. Polosa, JHEP **0307** (2003) 001; F. Caravaglios, M. L. Mangano, M. Moretti and R. Pittau, Nucl.Phys. **B 539** (1999) 232.
58. St. Frixione, et al., JHEP **06**(2002)029 .
59. C. F. Berger, Z. Bern, L. J. Dixon, F. Febres Cordero, D. Forde, T. Gleisberg, H. Ita, D. A. Kosower and D. Maître, *Next-to-Leading Order QCD Predictions for W +3-Jet Distributions at Hadron Colliders*, arXiv:0907.1984; Phys. Rev. Lett. **102** (2009) 222001, arXiv:0902.2760;
60. M. Spira and P. M. Zerwas, Lect. Notes Phys. **512** (1998) 161, hep-ph/9803257v2.
61. S. Moch and A. Vogt, Phys. Lett. **B 631** (2005) 48.
62. Ch. Anastasiou, K. Melnikov and F. Petriello, Nucl. Phys. **B 724** (2005) 197.
63. R. Harlander, Pramana **67** (2006) 875; hep-ph/0606095v1.
64. T. Sjöstrand, P. Edén, C. Friberg, L. Lönnblad, G. Miu, S. Mrenna and E. Norrbin, Comp. Phys. Comm. **135** (2001) 238; For colour reconnection and Bose-Einstein correlation studies PYTHIA version 6.121 is used. T. Sjöstrand, *Recent Progress in PYTHIA*, Preprint, LU-TP-99-42, hep-ph/0001032.
65. G. Davatz, et al., JHEP **07** (2006) 037.
66. R. M. Godbole and S. D. Rindani, Z. Phys. **C 36** (1987) 395.
67. K. Arnold, et al., KA-TP-31-2008, SFB/CPP-08-95, arXiv:0811.4559v2.
68. D. Zeppenfeld, *Collider Physics*, hep-ph/9902307.
69. O. Brein, M. L. Ciccolini, S. Dittmaier, A. Djouadi, M. Krämer and R. Harlander, hep-ph/0402003.
70. W. Beenakker, S. Dittmaier, M. Krämer, B. Plümper, M. Spira and P. M. Zerwas, Phys. Rev. Lett. **87** (2001) 201805; ibid. Nucl. Phys. **B 653** (2003) 151.
71. G. Corcella, I. G. Knowles, G. Marchesini, S. Moretti, K. Odagiri, P. Richardson, M. H. Seymour and B. R. Webber, JHEP **0101** (2001) 010; hep-ph/0210213.
72. LHADPDF, http://hepforge.cedar.ac.uk/lhapdf/; M. R. Whalley, D. Bourilkov, R. C. Group, hep-ph/0508110.
73. J. Pumplin, D. R. Stump, J. Huston, H. L. Lai, P. Nadolsky and W. K. Tung, JHEP **0207** (2002) 012, hep-ph/0201195; J. Pumplin, et al., *Parton Distributions and the Strong Coupling Strength: CTEQ6AB PDFs*, hep-ph/0512167.
74. A. D. Martin, W. J. Stirling, R. S. Thorne and G. Watt, *Update of Parton Distributions at NNLO*, arXiv:0706.0459v3.
75. J. M. Butterworth, J. R. Forshaw and M. H. Seymour *Multiparton Interactions in Photoproduction at HERA*, CERN-TH/95-82, MC-TH-96/05, UCL-HEP 96-02, hep-ph/9601371, Z. Phys. **C 72** (1996) 637; L. Lönnblad, M. Seymour, et al., *gamma-gamma Event Generators*, hep-ph/9512371; J. M. Butterworth and J. R. Forshaw, *Photoproduction of Multi-Jet Events at HERA - A Monte Carlo Simulation*, J.Phys. **G 19** (1993) 1657.
76. J. Fujimoto, et al., Comp. Phys. Comm. **100** (1997) 128; hep-ph/9605312.
77. F. A. Berends, R. Kleiss and R. Pittau, Nucl. Phys. **B 424** (1994) 308; Nucl. Phys. **B 426** (1994) 344; Nucl. Phys. (Proc. Suppl.) **B 37** (1994) 163; R. Kleiss and R. Pittau, Comp. Phys. Comm. **83** (1994) 141; R. Pittau, Phys. Lett. **B 335** (1994) 490.
78. S. Jadach, W. Placzek and B. F. L. Ward, Phys. Rev. **D 56** (1997) 6939.

79. E. Accomando, A. Ballestrero, E. Maina, Comput. Phys. Commun. **150** (2003) 166; hep-ph/0204052.
80. S. Jadach, B. F. L. Ward and Z. Wąs, Comp. Phys. Comm. **79** (1994) 503.
81. S. Jadach, B. F. L. Ward and Z. Wąs, *The precision Monte Carlo event generator KK for two-fermion final states in* e^+e^- *collisions*, CERN-TH/99-235.
82. S. Jadach, W. Placzek, B. F. L. Ward, Phys. Lett. **B 390** (1997) 298–308.
83. D. Karlen, Nucl. Phys. **B 289** (1987) 23.
84. R. Engel, Z. Phys. **C 66** (1995) 203; R. Engel and J. Ranft, Phys. Rev. **D 54** (1996) 4244.
85. S. Nova, A. Olshevski and T. Todorov, *MONTE-CARLO Event Generator for two Photon Physics*, DELPHI internal note 90-35, 1990.
86. F. A. Berends, P. H. Daverfeldt and R. Kleiss, Nucl. Phys. **B 253** (1985) 441.
87. J. A. M. Vermaseren, J. Smith and G. Grammer Jr, Phys. Rev. **D 19** (1979) 137; J. A. M. Vermaseren, Nucl. Phys. **B 229** (1983) 347.
88. E. Barberio and Z. Wąs, Comp. Phys. Comm. **79** (1994) 291.
89. Z. Was, Nucl. Phys. Proc. Suppl. **98** (2001) 96.
90. J. Alwall, et al., JHEP **09** (2007) 028.
91. B. P. Kersevan, E. Richter-Was, *The Monte Carlo Event Generator AcerMC version 3.4*, TPJU-6/2004, hep-ph/0405247.
92. K. Melnikov and F. Petriello, Phys. Rev. Lett. **96** (2006) 231803.
93. J. Campbell and R. K. Ellis, User Guide available at http://mcfm.fnal.gov/ (2007); J. Campbell, R. K. Ellis, G. Zanderighi, JHEP **0610** (2006) 028; J. Campbell, R. K. Ellis, F. Maltoni, S. Willenbrock, Phys. Rev. **D 73** (2006) 054007; J. Campbell, R. K. Ellis, F. Maltoni, S. Willenbrock, Phys. Rev. **D 75** (2007) 054015.
94. C. Balazs, E. Berger, P. Nadolsky and C. -P. Yuan, Phys. Lett. **B 637** (2006) 235.
95. L. Lönnblad, Comp. Phys. Comm. **71** (1992) 15.
96. See R. Brun et al, *GEANT 3*, CERN DD/EE/84-1 (Revised), September 1987; The GHEISHA program (H. Fesefeldt, RWTH Aachen Report PITHA 85/02, 1985) is used to simulate hadronic interactions.
97. S. Agostinelli, et al., Nucl. Inst. Meth. **A 506** (2003) 250; J. Allison, et al., IEEE Trans. Nucl. Sci. **53** (2006) 270.
98. Fast detector simulations are for example developed for the ATLAS detector: K. Mahboubi and K. Jakobs, *A fast parameterization of electromagnetic and hadronic calorimeter showers*, ATLAS Public Note ATL-SOFT-PUB-2006-001; D. Cavalli, et al., *Performance of the ATLAS fast simulation ATLFAST* ATLAS Internal Notes ATL-PHYS-INT-2007-005, ATL-COM-PHYS-2007-012; E. Barberio and A. Straessner, *Parameterization of electromagnetic showers for the ATLAS detector*, ATLAS Note ATL-COM-PHYS-2004-015.

Chapter 2
The LEP Experiments

The LEP experiments ALEPH, DELPHI, L3, and OPAL measured e^+e^- collisions from Z peak energies between 89 and 93 GeV up to highest energies above the W-pair threshold between 161 and 209 GeV. The goal of their experimental program was the determination of the properties of W and Z bosons, like their mass, width and their couplings to fermions and gauge bosons. It was also hoped to discover new phenomena, like the discovery of the Standard Model Higgs boson or supersymmetric particles. This was, however, not achieved. This chapter introduces the LEP collider and the important aspect of the calibration of the collision energy. The main features of the LEP experiments are presented.

2.1 The LEP Collider

The LEP ring [1] at CERN was installed in a tunnel of 26.7 km circumference at 50–175 m under ground and crossing the Swiss–French border. It was composed of eight 2.9 km long arc sections and eight 210 m long straight sections. The accelerator lattice was made of focusing-defocusing quadrupole and dipole structures, so-called FODO elements. Each of the element was 79 m long and 31 elements were arranged into one octant. The magnet system was built of 3,368 bending dipoles, together with about 800 quadrupoles for focusing and defocusing, and 500 sextupoles and further 600 dipoles for orbit correction. A bending field of up to 0.134 T created by steel-concrete dipoles kept the electrons circulating in the LEP ring with an effective bending radius of 3,026 m.

The acceleration of the electrons and positrons started with the 600 MeV linear LINAC injector for LEP (LIL). Both particle types were accumulated in the Electron-Positron-Accumulator (EPA) and the particle bunches were further accelerated in the Proton Synchrotron (PS) and the Super-Proton-Synchrotron (SPS) before they are eventually injected with an energy of about 20 GeV into LEP. The bunches collided every 22 μs at the interaction points (IP), where the experiments were installed. A collimator system protected the installation from synchrotron radiation. The CERN accelerator complex is shown in Fig. 2.1. The number of

A. Straessner, *Electroweak Physics at LEP and LHC*, STMP 235, 45–54,
DOI 10.1007/978-3-642-05169-2_2, © Springer-Verlag Berlin Heidelberg 2010

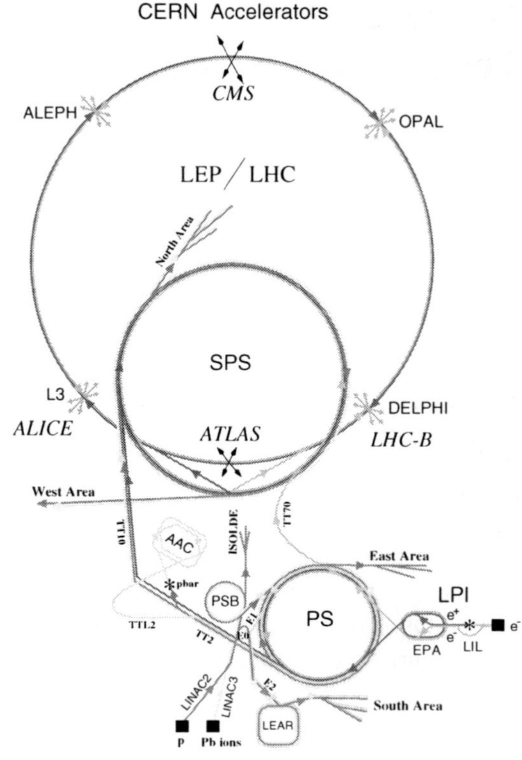

Fig. 2.1 The CERN accelerator complex with the various components for injection, storage and pre-acceleration of electrons, positrons, protons and heavy ions, and the LEP/LHC ring. The LEP experiments ALEPH, DELPHI, L3 and OPAL were installed in the underground caverns of IP 2, 4, 6, and 8. The LHC detectors ATLAS, CMS, LHC-b and ALICE are in IP 1, 5, 2, and 8, respectively

LEP: Large Electron Positron collider
SPS: Super Proton Synchrotron
AAC: Antiproton Accumulator Complex
ISOLDE: Isotope Separator OnLine DEvice
PSB: Proton Synchrotron Booster
PS: Proton Synchrotron
LPI: Lep Pre-Injector
EPA: Electron Positron Accumulator
LIL: Lep Injector Linac
LINAC: LINear ACcelerator
LEAR: Low Energy Antiproton Ring

interactions at the IPs is proportional to the luminosity of the e^+ and e^- beams, which is given by

$$L = \frac{N_b^2 n_b f_{rev}}{4\pi(\sigma^*)^2} , \tag{2.1}$$

where $N_b \approx 10^{11}$ is the number of particles per bunch, n_b the number of bunches per beam, $f_{rev} = 11{,}246\,s^{-1}$ the revolution frequency and $(\sigma^*)^2$ the transverse intersecting beam area at the IP. The luminosity is limited by electromagnetic beam–beam interactions between electron and positron bunches. They lead to a shift of the tune value Q, which describes the number of betatron oscillations per turn. For LEP, Q varied between 60 and 100, depending on the beam optics. The tune shift ΔQ had to be kept below 0.04 to provide stable running and optimal luminosity.

The maximisation of luminosity was achieved by increasing the number of bunches, n_b. In the first years, there were 4 bunches per beam, which was then changed to the "Pretzel" scheme with 8 equidistant bunches. Eventually, in the last years of LEP, a bunch train scheme with four trains of three bunches and later four trains of two bunches was employed.

The transverse betatron amplitude, σ^*, is also an important luminosity parameter. It is usually expressed in terms of the transverse emittance ε and the β-function at the IP, β^*:

$$\sigma^* = \sqrt{\varepsilon\beta^*} \tag{2.2}$$

In the horizontal plane, β^* reached 1.25 m mainly determined by horizontal oscillation damping due to emission of synchrotron radiation. In the vertical plane β^* was about 4 cm. Both values were achieved by installing strong focusing superconducting quadrupoles with a high gradient of 55 T/m at the IPs.

The final luminosity achieved was 4.3×10^{31} cm^{-2}s^{-1} at 46 GeV beam energy and about 10^{32} cm^{-2}s^{-1} at 100 GeV, with a beam current of 1 mA per bunch.

The record beam energy of 104.5 GeV that could be reached was limited by synchrotron radiation. The energy loss per turn is in good approximation given by

$$\Delta E = \frac{4\pi\alpha_{\rm QED}}{3}\frac{1}{m_e^4}\frac{E_b^4}{R}\,, \tag{2.3}$$

with the electron mass, m_e, the beam energy, E_b, and the effective bending radius, $R \approx 3026\,m$. At $E_b = 104.5$ GeV the loss was therefore about 3.3% of the beam energy per turn, which had to be compensated by the accelerating radio-frequency (RF) power. The RF cavity system [2] was installed in the straight sections. For Z pole energies in the LEP1 phase, 128 five-cell copper cavities were sufficient to supply the acceleration power. For high energy operation in the LEP2 phase, the cavities were replaced by 288 superconducting four-cell cavities running at 352 MHz, 31,320 times the revolution frequency, f_{rev}. To reach the highest energies 56 copper cavities were added to finally achieve a total voltage of 3,630 MV, corresponding to an average gradient of 7.5 MV/m. This dramatically exceeded the original cavity design value of 6 MV/m and was only possible by special cavity conditioning.

2.2 LEP Energy Calibration

The calibration of the beam energy [3, 4] was of primordial importance during both LEP phases to determine $M_{\rm Z}$ and $M_{\rm W}$ with high precision. At LEP1, the Z boson mass was derived from the measurement of the fermion-pair cross-sections mainly at the 91.2 GeV peak of the resonance shape and at two off-peak points ± 1.8 GeV above and below the peak. The contribution of the LEP energy uncertainty to the Z mass and width error is approximatively given by [13]:

$$\Delta M_Z \approx 0.5\Delta(E_{p+2} + E_{p-2}) \tag{2.4}$$

$$\Delta\Gamma_Z \approx \frac{\Gamma_Z}{E_{p+2} - E_{p-2}}\Delta(E_{p+2} - E_{p-2}) \ , \tag{2.5}$$

where E_{p-2} and E_{p+2} are the two off-peak centre-of-mass energies.

The best method to measure the beam energy is by resonant depolarisation. The Sokolov–Ternov effect [5, 6] provides the mechanism for transverse polarisation of the beam electrons. Due to synchrotron radiation the electron spin is aligned in the magnetic dipole field. The degree of polarisation is measured with a Compton polarimeter using polarised laser light. The beam polarisation is disturbed by a transverse oscillating magnetic field of a certain frequency, ν_r. Depolarisation resonance appears if the ratio ν_r/ν_{rev} is equal to the non-integer part of the number of spin precessions per turn, also called spin tune, ν_s. For the three energy scan points, E_{p-2}, E_p, E_{p+2}, the spin tunes were 101.5, 103.5 and 105.5 respectively. The corresponding beam energy can then be determined from the relation

$$E_b = \frac{\nu_s m_e}{(g_e - 2)/2} \ , \tag{2.6}$$

where m_e is the electron mass and $(g_e - 2)/2$ the anomalous magnetic moment of the electron. The resonant depolarisation method yields a beam energy precision below 1 MeV. The main measurements were performed on the electron beam only, but a few cross-calibration measurements with the positron beam showed that both beam energies agree well within less than 0.4 MeV.

Sufficient beam polarisation could however only be achieved for E_b up to 61 GeV. Also, the energy calibration could only be performed in dedicated calibration runs and not during physics data taking. The precise energy values had therefore to be extrapolated to physics runs and to other beam energies. This is performed by means of the strength of the magnetic dipole field B which, after integration over the whole LEP ring, is proportional to the beam energy

$$E_b = \frac{ec}{2\pi}\oint B\, ds \ . \tag{2.7}$$

A continuous measurement of the B field was therefore performed using nuclear magnetic resonance (NMR) probes installed inside the magnets, as shown in Fig. 2.2. In the LEP1 phase, 4 NMR probes were read out and 12 more probes were added for LEP2. A LEP energy model [4, 7] was developed to derive the beam energy from the NMR measurements. Many time-dependent details were taken into account, for example the variation of the bending field due to parasitic currents flowing along the beam pipe (the "TGV effect"), the monitored dipole temperature, corrections due to tidal movement of the LEP ring (the "moon effect"), as well as corrections due to the beam orbit position.

Three further and complementary measurement methods were applied to estimate the uncertainty of the reference energy determination with NMR. A magnetic

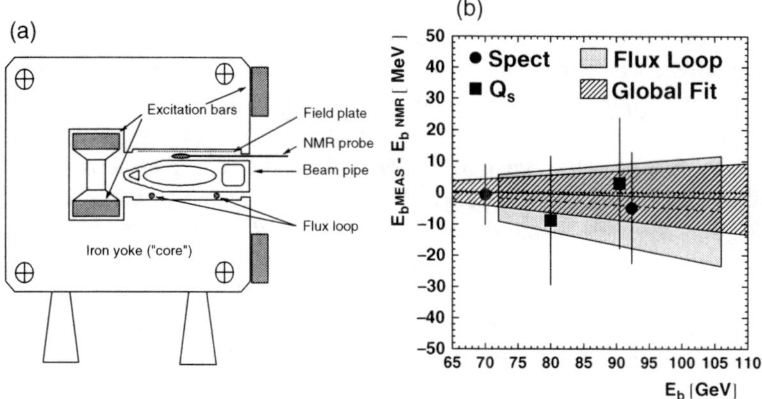

Fig. 2.2 (**a**) Schematic view of the LEP reference dipole magnet with NMR probe and flux loop installed. (**b**) Comparison of the beam energy measurements using the LEP spectrometer, the flux loop and the tune shift, Q_s, with the measurement by the NMR probes. Also shown is the result of the global calibration fit [4]

flux-loop was installed in one special dipole magnet. It determined the magnetic field induced in a large copper loop during the ramping of the magnet currents. In the last year of LEP running, a beam spectrometer made of a steel dipole and a triplet of beam-position monitors provided a second alternative energy measurement. Finally, the beam energy can be determined from the synchrotron tune, Q_s. It is defined as the ratio of the longitudinal beam oscillation frequency to the revolution frequency. The longitudinal beam oscillations are a combined effect of energy loss due to synchrotron radiation and the acceleration in the RF fields. From the relative phase of a bunch and the RF voltage Q_s was measured and, knowing the RF peak voltage, E_b could be calculated. Figure 2.2 compares the three alternative methods to the nominal NMR energy calibration as a function of beam energy. The methods yield consistent results.

The systematic uncertainties in the final calibration originated mainly from this comparison, which contributes with about 20 MeV to the uncertainty on the centre-of-mass energy at LEP2. Additional 10 MeV are from the modelling of the energy loss between the RF stations, which is needed to determine the exact energy at each IP and also to relate the LEP spectrometer measurements to those of the NMR method. A special situation was in the last year of LEP running, where previously unused horizontal correction dipoles were used. With these additional magnets the bending field was spread over a longer trajectory which leads to an increase in attainable beam energy. The gain was about 120 MeV per beam, which was pushing the discovery potential of LEP to higher particle masses. The downside was an increased systematic uncertainty of about 30 MeV on $\sqrt{s}$, however, only for the highest centre-of-mass energies $\sqrt{s} > 205$ GeV. Sources of smaller systematic uncertainties were also studied like the e^+e^- energy difference, the beam energy variation during the fill, the variation of the RF frequency, the precision of the resonant depolarisation measurement, and the additional dipole field component due an imbalance in the current feeding the focusing and defocusing quadrupoles.

Eventually, an IP-dependent calibrated centre-of-mass energy was provided in time steps of 15 minutes. A precision of 2–3 MeV and 3–7 MeV was reached for each off-peak and on-peak point of the Z resonance scan, respectively. The beam energy spread was in the order of 55 MeV. At higher energies above the W pair threshold, the centre-of-mass energy was calibrated for most of the energy points to better than 25 MeV, while the beam energy spread was about 250 MeV. Since the calibration procedure applies common corrections for the energy points, also correlations are determined and taken into account. The very good understanding of the LEP accelerator is eventually the basis for the precise measurements of mass and width of the Z and W bosons performed by the LEP experiments.

2.3 The ALEPH, DELPHI, L3 and OPAL Experiments

The four LEP detectors, ALEPH [8], DELPHI [9], L3 [10] and OPAL [11], are multi-purpose detectors designed to measure the products of head-on e^+e^- collisions in their centre. A schematic view of the different sub-detector systems installed in the four experiments is shown in Figs. 2.3 and 2.4. The experiments are all equipped with silicon tracking detectors close to the interaction point. The silicon devices are arranged cylindrically around the beam pipe, typically at radii between 5 and 15 cm. Their main purpose is to resolve secondary vertices from B hadron decays, which travel about 3 mm before decaying. With an impact parameter resolution below $100\,\mu$m, b quark decays of the Z boson can be separated from light quark decays, and b decay modes of a possible Higgs boson can be identified. The silicon detectors are surrounded by gas drift chambers, where different technologies are used. ALEPH and OPAL installed a tracking or vertex chamber at smaller radii, completed by a time projection or jet chamber used for tracking of charged particles at larger radii up to about 2 m. L3 had a single time expansion chamber [12] with an outer radius of 60 cm, while DELPHI used a time projection chamber. Identification of particles was done by determination of ionisation energy loss along the tracks, dE/dx. DELPHI used a Ring Image Cherenkov detector for separating relativistic particles of different mass. For the measurement of track momenta, solenoids provide a magnetic bending field between 0.5 and 1.5 T, which covers at least the inner tracking detectors.

The measurement of the energy of electromagnetic particles, like photons and electrons, is performed by electromagnetic calorimeters. The ALEPH detector used a lead/wire chamber sampling technique, while lead glass and bismuth germanate (BGO) crystals were installed in OPAL and L3, respectively. DELPHI used a high density projection chamber with lead absorber walls for electromagnetic calorimetry. Sufficient material density of in the order of 20 radiation lengths, X_0, guaranteed that the electromagnetic showers and energy depositions are contained in the calorimeters.

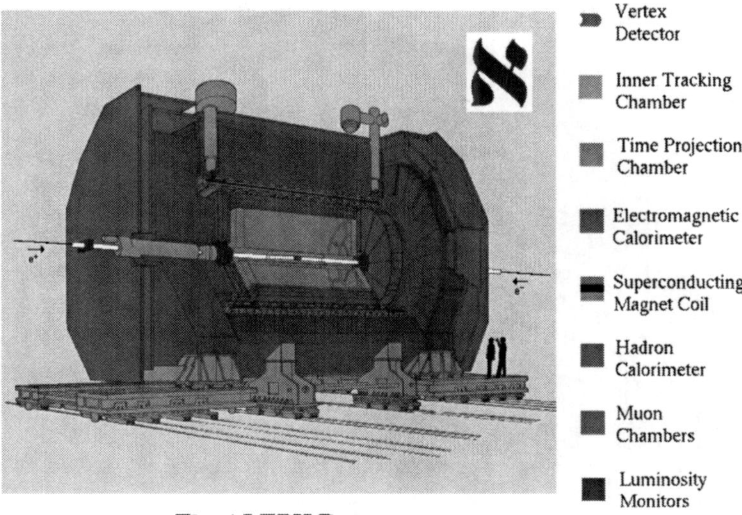

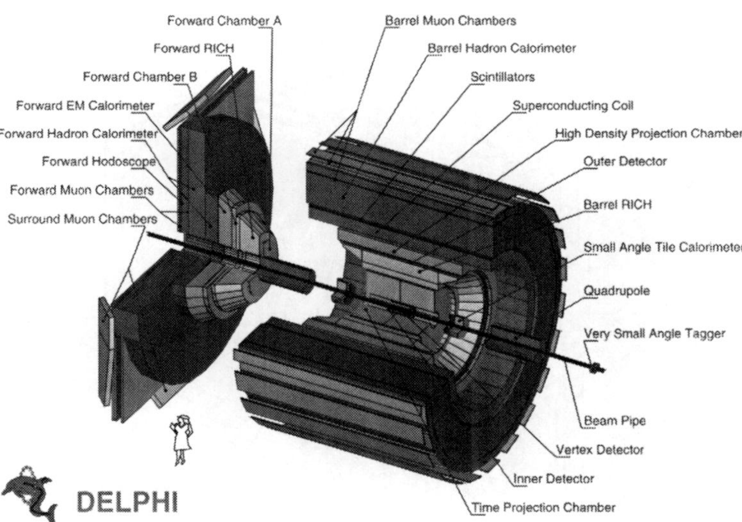

Fig. 2.3 The ALEPH and DELPHI detectors at the LEP collider

Jets from fragmented quarks and gluons usually traverse the electromagnetic detectors and were registered and eventually stopped in the hadronic calorimeters. Here, ALEPH, DELPHI and OPAL used the magnetic return yoke made of iron as absorber, equipped with streamer chambers or tubes. L3 had a depleted uranium and wire chamber sampling calorimeter. The minimal ionising muons are not stopped in the inner detector and calorimeter layers and were measured in a muon detection system in the outermost shell of the LEP experiments. The correct event timing

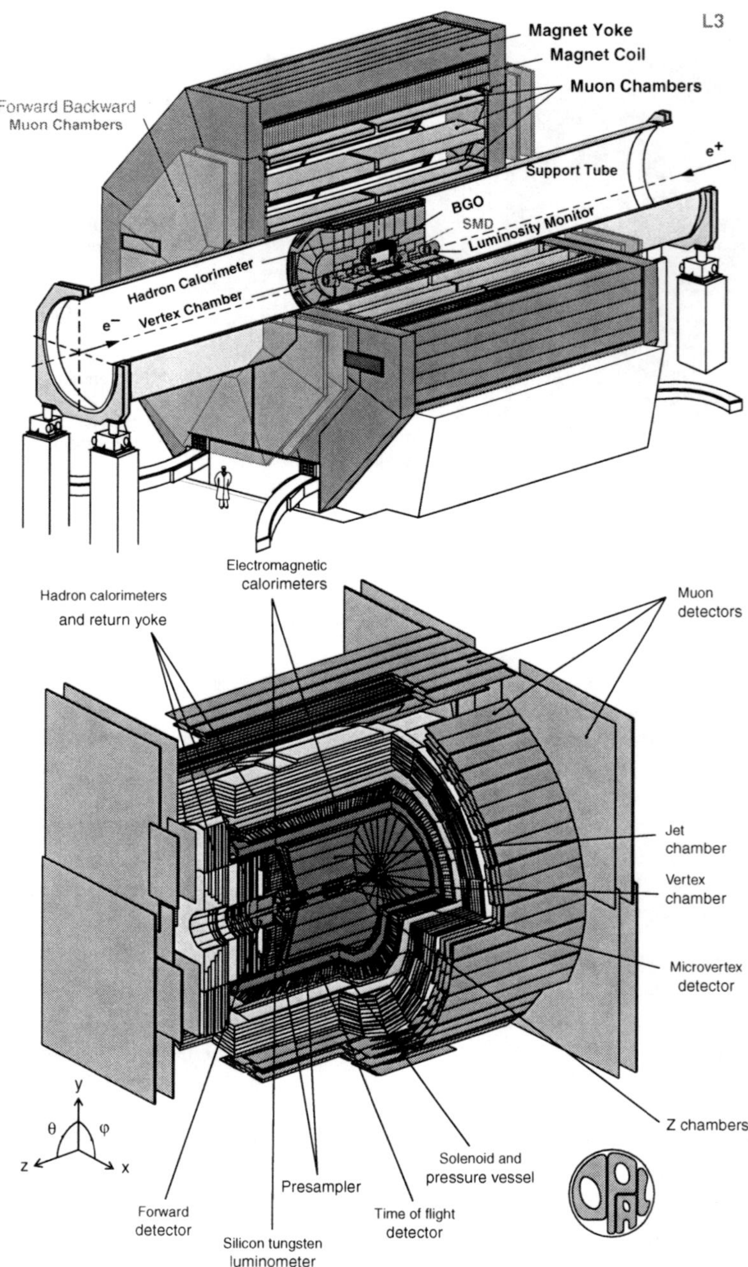

Fig. 2.4 Schematic view of the L3 and OPAL detectors

and rejection of cosmic ray background was performed by scintillator time-of-flight systems.

The tracking performance for muons from Z peak decays, $Z \rightarrow \mu^+\mu^-$, reaches a resolution between 1.5 and 2.5% for the LEP experiments. Electrons and photons are measured with about 1–2.5% energy resolution. The uncertainty on hadronic jet energies is in the order of 10% for 45 GeV jets.

Important for cross-section measurements of the different physics processes is the precise knowledge of the beam luminosity, L. At LEP, small-angle Bhabha scattering served as a reference process to determine L. This process is well described by QED and has small electroweak corrections when the acceptance region is restricted to small polar angles between θ_{min} and θ_{max}. At lowest order the differential cross-section at small scattering angles is given by

$$\frac{d\sigma}{d\Omega} = \frac{\alpha^2}{s}\frac{1}{\sin^4(\theta/2)} \approx \frac{16\alpha^2}{s}\frac{1}{\theta^4} . \tag{2.8}$$

Integrating over the acceptance angles and using $d\Omega \approx 2\pi\theta\, d\theta$, yields

$$\sigma_{acc} = \frac{16\pi\alpha^2}{s}\left(\frac{1}{\theta_{min}^2} - \frac{1}{\theta_{max}^2}\right) . \tag{2.9}$$

The electrons and positrons of the small-angle Bhabha process are detected in luminosity monitors installed in the very forward regions of the detectors. To obtain the 0.1% precision on the luminosity, the fiducial volumes have to be very well defined. Therefore a combination of electromagnetic calorimetry and silicon devices for exact angular measurement are used. The final uncertainties actually fully reached the expectations, and the dominating systematic effects on Z peak cross-section measurements were due to the limited precision of the theoretical prediction for the Bhabha cross-section in the angular range [13].

References

1. R. Bailey, C. Benvenuti, S. Myers and D. Treille, C. R. Physique **3** (2002) 1107–1120, and references therein.
2. J. P. H. Sladen, *The RF System to 102 GeV : How did we get there and can we go further ?*, Proceedings of the 10th Chamonix Workshop on LEP-SPS, Ed. P. Le Roux, et al., CERN-SL-2000-007.
3. The LEP Energy Working Group, A. Blondel et al., Eur. Phys. J. **C 11** (1999) 573.
4. LEP Energy Working Group, R. Assmann, et al., Eur. Phys. J. **C 39** (2005) 253.
5. M. Sokolov and I. M. Ternov, Dokl. Akad. Nauk SSSR **153** (1963) 1053; Sov. Phys. Dokl. **8** (1964) 1203.
6. A. A. Sokolov, I. M. Ternov and Yu. M. Loskutov, *Effect of Quantum Fluctuation on Vertical Oscillations of an Electron Moving in a Magnetic Field*, Phys. Rev. **125** (1962) 731.
7. The LEP Energy Working Group, R. Assmann et al., Eur. Phys. J. **C 6** (1999) 187–223.
8. The ALEPH Collaboration, Nucl. Inst. Meth. **A 294** (1990) 121; Nucl. Inst. Meth. **A 360** (1995) 481.

9. The DELPHI Collaboration, P. Aarnio et al., Nucl. Instr. Meth. **A 303** (1991) 233; P. Abreu et al., Nucl. Instr. Meth. **A 378** (1996) 57; The DELPHI Silicon Tracker Group, P. Chochula et al., Nucl. Instr. Meth. **A 412** (1998) 304; S. J. Alvsvaag et al., Nucl. Instr. Meth. **A 425** (1999) 106.
10. The L3 Collaboration, B. Adeva et al., Nucl. Instr. Meth. **A 289** (1990) 35; M. Chemarin et al., Nucl. Instr. Meth. **A 349** (1994) 345; M. Acciarri et al., Nucl. Instr. Meth. **A 351** (1994) 300; G. Basti et al., Nucl. Instr. Meth. **A 374** (1996) 293; I.C. Brock et al., Nucl. Instr. Meth. **A 381** (1996) 236; A. Adam et al., Nucl. Instr. Meth. **A 383** (1996) 342.
11. The OPAL Collaboration, K. Ahmet et al., Nucl. Instr. Meth. **A 305** (1991) 275; B. E. Anderson et al., IEEE Trans. Nucl. Sci. **41** (1994) 845; S. Anderson et al., Nucl. Instr. Meth. **A 403** (1998) 326.
12. H. Anderhub et al., Nucl. Instr. Meth. **A 515** (2003) 31.
13. The ALEPH Collaboration, the DELPHI Collaboration, the L3 Collaboration, the OPAL Collaboration, the SLD Collaboration, the LEP Electroweak Working Group, the SLD electroweak, heavy flavour groups, Phys. Rept. **427** (2006) 257; hep-ex/0509008v3.

Chapter 3
Gauge Boson Production at LEP

The main research goals of the LEP program were the detailed study of the Z and W boson properties. In the first phase of LEP, the Z line-shape was explored to exactly determine the Z mass, width and the Z-fermion couplings. The data collected by the four LEP experiments at energies around the Z peak consists of 17 million Z decays, $e^+e^- \to Z \to f\bar{f}(\gamma)$, completed by 600 thousand Z decays measured by the SLD experiment at the SLC. A short summary of the main results shall be given here.

At energies above the Z peak, single-Z production, $e^+e^- \to Ze^+e^-$, and Z-pair production, $e^+e^- \to ZZ$, are kinematically accessible. The reaction $e^+e^- \to Z+\gamma$ was studied as a possible calibration process for the LEP beam energy. This calibration is important for the determination of the mass of the other heavy gauge boson, the W.

The massless photon is ubiquitous [1] in all reactions as it is radiated by charged particles and therefore usually included in the definition of the physics process, like in $e^+e^- \to Z \to f\bar{f}(\gamma)$, for example. More interesting are the non-inclusive processes, like Compton scattering, $e^+e^- \to \gamma e^+e^-$ and photon pair-production $e^+e^- \to \gamma\gamma(\gamma)$.

At LEP energies above 161 GeV, W bosons are produced in pairs, $e^+e^- \to WW$ and singly in the process $e^+e^- \to We\nu$. The measurement of the corresponding cross-sections gives insight into the non-abelian structure of the boson couplings in the Standard Model. Pair production probes the $WW\gamma$ and WWZ vertex, while the single-W process involves only the $WW\gamma$ vertex. In the following, the measurements at the LEP collider are described.

3.1 Z Pole Measurements at LEP and SLD

At centre-of-mass energies around the Z pole the properties of the Z boson were determined with high precision. At LEP, the Z mass and width were derived from the line-shape of the Z resonance [2]. Forward-backward asymmetries of the Z decay products as well as decay branching fractions were determined by LEP and SLD for leptons, hadrons, and also separately for heavy quarks flavours. The polarised beam of the SLC allowed a measurement of the left-right asymmetry in leptonic Z decays.

A. Straessner, *Electroweak Physics at LEP and LHC*, STMP 235, 55–110,
DOI 10.1007/978-3-642-05169-2_3,

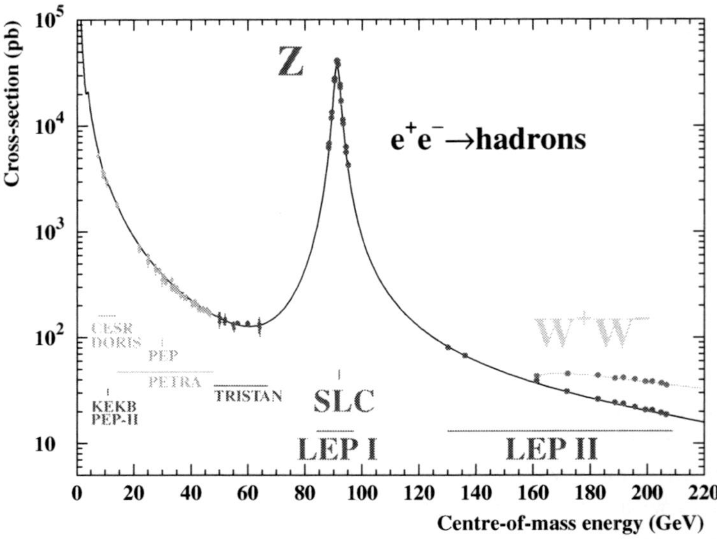

Fig. 3.1 Measured and predicted hadronic cross-section in e^+e^- collisions as a function of centre-of-mass energy [2]

To measure the mass and width of the Z boson, the e^+e^- centre-of-mass energy of LEP was varied over a range of ± 3 GeV around $\sqrt{s} = M_Z$. At the different scan points the hadronic and leptonic cross-sections were measured to derive the resonance curve. The evolution of the hadronic cross-section in e^+e^- collisions is illustrated in Fig. 3.1.

By scanning the line-shape of the Z resonance, the mass and width of the Z boson were determined by the four LEP experiments to be [2]

$$M_Z = 91.1875 \pm 0.0021 \text{ GeV} \tag{3.1}$$

$$\Gamma_Z = 2.4952 \pm 0.0023 \text{ GeV} \tag{3.2}$$

assuming lepton universality. The precision of the Z mass is at the ppm level and is comparable to that of the muon decay constant G_F [3].

Further properties of the Z boson describe the production and decay at the resonance peak. They are usually summarised in a few observables that are input to the global electroweak analysis. The hadronic peak cross-section

$$\sigma^0_{\text{had}} = \frac{12\pi}{M_Z^2}\frac{\Gamma_{\text{ee}}\Gamma_{\text{had}}}{\Gamma_Z^2} = \frac{12\pi}{M_Z^2}\frac{\Gamma_{\text{ee}}\sum_{q\neq t}\Gamma_{q\bar{q}}}{\Gamma_Z^2} \tag{3.3}$$

is measured as [2]:

$$\sigma^0_{\text{had}} = 41.450 \pm 0.037 \text{ nb.} \tag{3.4}$$

The analysis of leptonic decays into e^+e^-, $\mu^+\mu^-$ and $\tau^+\tau^-$ pairs yields the hadronic to leptonic decay width:

$$R_\ell^0 = \frac{\Gamma_{\text{had}}}{\Gamma_{\ell\ell}} = 20.767 \pm 0.025, \tag{3.5}$$

assuming lepton universality, which is experimentally confirmed by the good agreement of the measurement of the individual ratios, R_e, R_μ, and R_τ [2]. Furthermore, the invisible Z decay width, $\Gamma_{\text{inv}} = \Gamma_Z - \Gamma_{\text{had}} - \Gamma_{ee} - \Gamma_{\mu\mu} - \Gamma_{\tau\tau}$, can be derived from R_ℓ^0 as:

$$R_{\text{inv}}^0 = \frac{\Gamma_{\text{inv}}}{\Gamma_{\ell\ell}} = \left(\frac{12\pi R_\ell^0}{\sigma_{\text{had}}^0 M_Z^2}\right)^{\frac{1}{2}} - R_\ell^0 - (3 + \delta_\tau), \tag{3.6}$$

where δ_τ is correcting for the mass of the τ lepton. From this ratio the number of light neutrino species, N_ν, is extracted assuming Standard Model couplings of the neutrinos to the Z boson:

$$R_{\text{inv}}^0 = N_\nu \left(\frac{\Gamma_{\nu\nu}}{\Gamma_{\ell\ell}}\right)_{\text{SM}}. \tag{3.7}$$

The analysis of the LEP data yields:

$$N_\nu = 2.9840 \pm 0.0082, \tag{3.8}$$

which is deviating from three by only 2 standard deviations. This is nicely illustrated in Fig. 3.2. A more direct determination of the invisible Z width is obtained from studies of single- and multi-photon events produced in the reaction

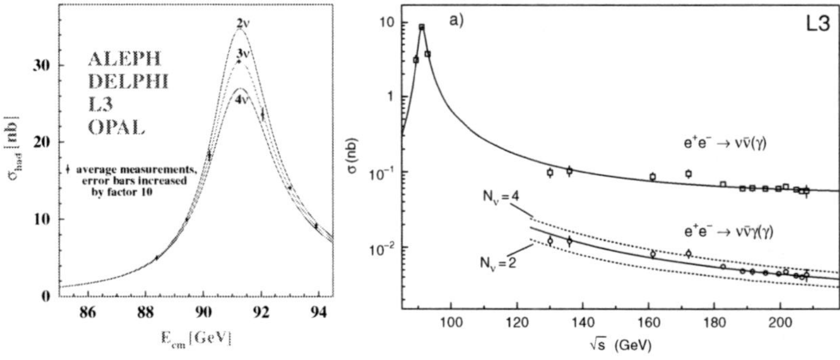

Fig. 3.2 (**a**) Measured hadronic cross-section at Z peak energies as a function of centre-of-mass energy [2], compared to predictions with 2, 3, and 4 light neutrino generations with Standard Model couplings. (**b**) Cross-sections of $e^+e^- \to \nu\bar{\nu}(\gamma)$ and $e^+e^- \to \nu\bar{\nu}\gamma(\gamma)$ processes. The latter is directly sensitive to the number of neutrino generations and scenarios with $N_\nu = 2, 3, 4$ are compared to L3 data [4]

$e^+e^- \rightarrow \nu\bar{\nu}\gamma(\gamma)$, which results in $N_\nu = 2.92 \pm 0.05$ [3, 4], again compatible with three light neutrino generations. The corresponding cross-section measurements from which this number is derived is shown in Fig. 3.2.

Additional information on the Z coupling structure to the fermions is contained in the asymmetry measurements. They depend on the helicity of the colliding electrons and positrons and on the polarisation of the produced particles. The forward-backward asymmetry is defined as

$$A_{\mathrm{FB}} = \frac{N_F - N_B}{N_F + N_B}, \tag{3.9}$$

where N_F (N_B) denotes the number of events in which the fermion is produced in the forward (backward) hemisphere, with polar angles $\theta < \pi/2$ ($\theta > \pi/2$) with respect to the incoming electron beam. The leptonic asymmetry at the Z pole is measured to be:

$$A_{\mathrm{FB}}^{0,\ell} = 0.0171 \pm 0.0010. \tag{3.10}$$

For τ leptons, also the polarisation can be determined by studying the kinematic of the observed τ decay products. For example, in $\tau \rightarrow \pi\nu_\tau$ decays, the energy spectrum of pions in the tau rest frame depends on the tau helicity. These measurements yield

$$\mathcal{A}_\ell(\mathcal{P}_\tau) = 0.1465 \pm 0.0033, \tag{3.11}$$

assuming universality of taus and electrons.

The leptonic data are completed by the measurement of the left-right asymmetry at SLD

$$A_{\mathrm{LR}} = \frac{N_L - N_R}{N_L + N_R} \frac{1}{\langle \mathcal{P}_\mathrm{e} \rangle}, \tag{3.12}$$

from which the leptonic asymmetry parameter is derived as

$$\mathcal{A}_\ell = 0.1514 \pm 0.0022, \tag{3.13}$$

using Eq. (1.79). After combination with the determination of the left-right forward-backward asymmetry, $A_{\mathrm{LR,FB}}$, the measured value of $\mathcal{A}_\ell$ is only slightly changed:

$$\mathcal{A}_\ell = 0.1513 \pm 0.0021. \tag{3.14}$$

The leptonic asymmetry measurements rely on correct charge tagging, which, in case of e, μ and τ, is rather precise, because the single-track charge confusion is usually small and elementary leptons have unit charge. This is however not the case for quarks and more refined methods are introduced to determine the $\mathrm{q\bar{q}}$ charge asymmetry in Z decays. By weighting the charged tracks in quarks jets according to their momentum, the forward and backward hemispheres can be assigned to the quark jets. The combined result of these measurements is expressed in terms of the effective leptonic weak mixing angle:

$$\sin^2\theta^{\ell}_{\mathrm{eff}}\left(Q^{\mathrm{had}}_{\mathrm{FB}}\right) = 0.2324 \pm 0.0010 \tag{3.15}$$

Identifying the heavy quark flavours yields additional information on the details of the Z decays. The combination of LEP and SLD data results in the following branching fractions and asymmetries for c and b quarks separately:

$$R^0_{\mathrm{b}} = 0.21629 \pm 0.00066 \tag{3.16}$$

$$R^0_{\mathrm{c}} = 0.1721 \pm 0.0030 \tag{3.17}$$

$$A^{0,\mathrm{b}}_{\mathrm{FB}} = 0.0992 \pm 0.0016 \tag{3.18}$$

$$A^{0,\mathrm{c}}_{\mathrm{FB}} = 0.0707 \pm 0.0035 \tag{3.19}$$

$$\mathcal{A}_{\mathrm{b}} = 0.923 \pm 0.020 \tag{3.20}$$

$$\mathcal{A}_{\mathrm{c}} = 0.670 \pm 0.027 \tag{3.21}$$

The asymmetry parameters, $\mathcal{A}_{\mathrm{f}}$, depend only on the ratio of the effective vector and axial-vector coupling constants, $g^{\mathrm{f}}_{\mathrm{V}}/g^{\mathrm{f}}_{\mathrm{A}}$, while the square-root of their squared sum enters the partial Z decay widths. At LEP and SLD, especially the leptonic coupling constants are determined with high precision:

$$g^{\nu}_{\mathrm{A}} = g^{\nu}_{\mathrm{V}} = +0.50076 \pm 0.00076 \tag{3.22}$$

$$g^{\ell}_{\mathrm{A}} = -0.50123 \pm 0.00026 \tag{3.23}$$

$$g^{\ell}_{\mathrm{V}} = -0.03783 \pm 0.00041 \tag{3.24}$$

with an anti-correlation of 48% between g^{ν}_{A} and g^{ℓ}_{A}, respectively g^{ν}_{V} and g^{ℓ}_{A}, and only small correlations around 5% between g^{ℓ}_{V} and the other couplings. A comparison between the individual results of the three leptons, e, μ and τ, shows the universality of the leptonic couplings which is illustrated in Fig. 3.3. The variation of the Standard Model prediction is also indicated, which agrees best with data if the Higgs boson is light. The value of g^{ℓ}_{A} differs from the tree level value of $T_3 = -\frac{1}{2}$ by 4.7 standard deviations, showing clearly the presence of electroweak radiative corrections.

The various asymmetry measurements allow an extraction of the effective weak mixing angle, $\sin^2\theta_{\mathrm{eff}}$, independent of the value of the ρ parameter, when exploiting Eq. (1.92), since only the ratio of the coupling constants, $g^{\mathrm{f}}_{\mathrm{V}}/g^{\mathrm{f}}_{\mathrm{A}}$, appears. The results are shown graphically in Fig. 3.3, where the measurements are compared in terms of the leptonic $\sin^2\theta^{\ell}_{\mathrm{eff}}$. The two groups of leptonic and hadronic measurements each show very good agreement, while the comparison between the two is somewhat less consistent, although not with a very strong significance. This may hint to deviation in the hadronic sector, which is enhanced when looking at the right-handed

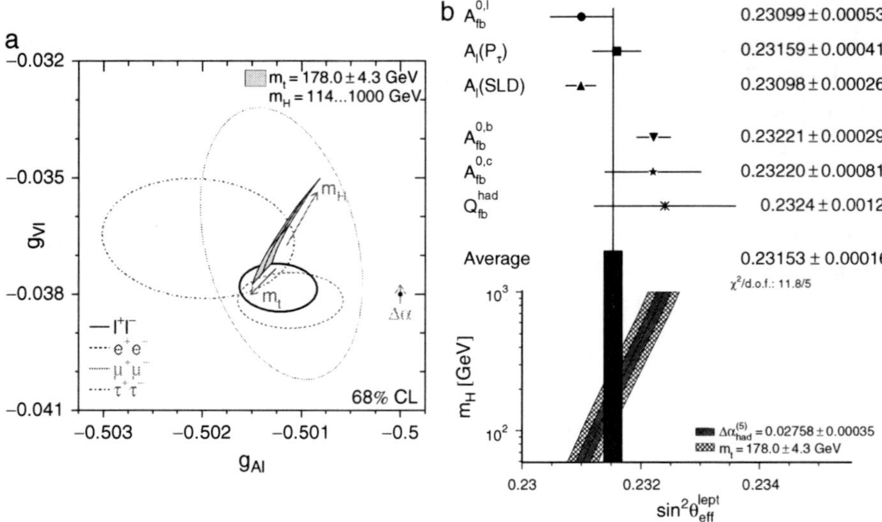

Fig. 3.3 (**a**) Effective vector and axial-vector coupling constants for leptons measured by the LEP experiments and SLC [2], compared to each other and to the Standard Model prediction for different values of m_t and M_H. The *arrow* indicates the variation of the theoretical prediction with $\Delta\alpha_{had}^{(5)}$. (**b**) Measurements of the leptonic effective electroweak mixing angle at LEP and SLD [8]

b-coupling, as done in Fig. 3.4. This will however remain an unanswered question until new data may become available, possibly from Z boson decays at the hadron colliders Tevatron and LHC or from an international linear collider (ILC). For the latter, the option of a high-luminosity running at the Z pole, called GIGAZ [5], is discussed.

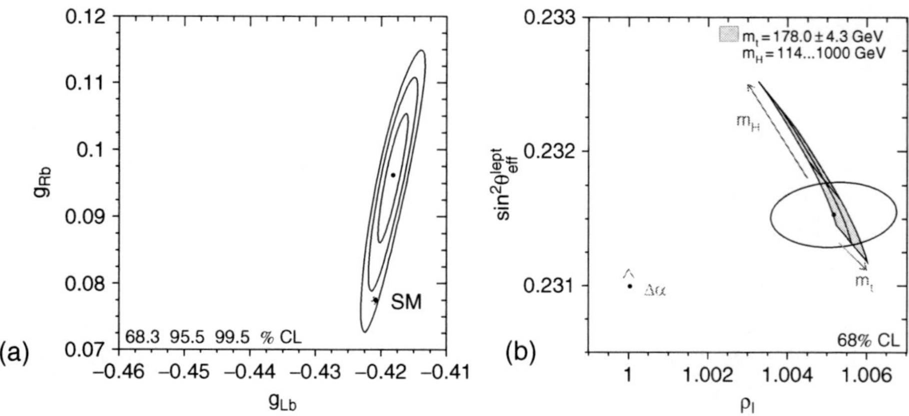

Fig. 3.4 (**a**) *Left* and *right* handed b-quark coupling derived from LEP and SLD data. (**b**) Contour curve of the ρ_ℓ and $\sin^2\theta_{eff}^{\ell}$ measurements compared to the Standard Model predictions where M_H, m_t and $\Delta\alpha_{had}^{(5)}$ are varied within the given ranges [2]

With the measured values of g_V^f and g_A^f by LEP and SLC the combined leptonic effective weak mixing angle and the ρ_ℓ parameter are extracted according to Eq. (1.53) [2], which yields:

$$\sin^2\theta_{\text{eff}}^{\ell} = 0.23153 \pm 0.00016 \tag{3.25}$$

and

$$\rho_\ell = 1.0050 \pm 0.0010. \tag{3.26}$$

The Standard Model predictions at tree level

$$\sin^2\theta_{\text{w}} = \frac{1}{2}\left(1 - \sqrt{1 - 4\frac{\pi\alpha_{\text{QED}}\left(M_Z^2\right)}{\sqrt{2}G_F M_Z^2}}\right) = 0.23098 \pm 0.00012 \tag{3.27}$$

$$\rho = 1 \tag{3.28}$$

deviate from these by 2.8 and 5.0 standard deviations. This is again a clear evidence for the existence of significant radiative corrections. Figure 3.4 shows the measured and predicted values in the ρ_ℓ-$\sin^2\theta_{\text{eff}}^{\ell}$ plane. It is interesting to note that best agreement between theory and experiment is achieved for light Higgs masses, as mentioned before, but also for top masses around 174 GeV, which agrees better with the recent m_t result [76] than with the slightly higher measurement used at the time the graphic was produced [2].

3.2 Neutral Boson Production Above the Z Peak

3.2.1 Photon Production

The QED part of the neutral boson sector is tested at low energy to high precision [6]. At LEP energies, pure QED processes are as well found to be in nice agreement with the theoretical predictions. Figure 3.5 documents the measurements of quasi-real Compton scattering $e^+e^-\to\gamma e^+e^-$. In this reaction, one beam electron (or positron) emits a quasi-real photon with low virtuality, $Q^2 < 2$ GeV, and escapes along the beam pipe. The photon and the beam positron (or electron) are scattered and measured in the detector. The effective centre-of-mass energy of this process can be calculated assuming three-particle kinematics:

$$\sqrt{s'} = \sqrt{1 - \frac{2E_{\text{miss}}}{\sqrt{s}}}\text{, with} \tag{3.29}$$

$$E_{\text{miss}} = \sqrt{s}\frac{|\sin(\theta_e + \theta_\gamma)|}{\sin\theta_e + \sin\theta_\gamma + |\sin(\theta_e + \theta_\gamma)|}. \tag{3.30}$$

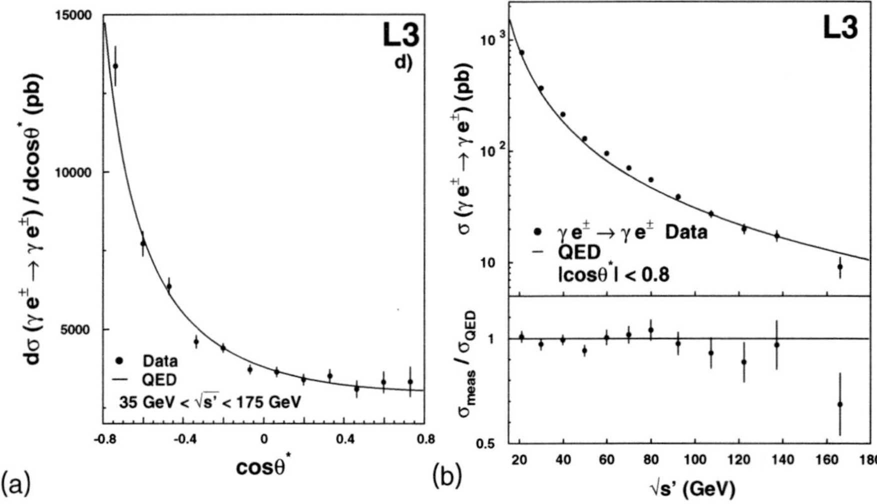

Fig. 3.5 Angular distribution of the detected electron and total cross-section of quasi-real Compton scattering, as measured by L3 [7]

To reduce background from Bhabha scattering the effective scattering angle, $\cos\theta^*$, is constrained to the central region

$$|\cos\theta^*| = \left|\frac{\sin(\theta_\gamma - \theta_e)}{\sin\theta_\gamma + \sin\theta_e}\right| < 0.8. \tag{3.31}$$

The typical backward scattering angular distribution, which is given to lowest order by

$$\frac{d\sigma}{d\cos\theta^*} = \frac{\alpha_{\text{QED}}^2\pi}{s'}\left(\frac{1+\cos\theta^*}{2} + \frac{2}{1+\cos\theta^*}\right), \tag{3.32}$$

is shown in Fig. 3.5. The LEP measurement is sensitive to very high values of $\sqrt{s'} = 175$ GeV, that were never reached before.

A similar good agreement with QED predictions is found for real photon production $e^+e^- \to \gamma\gamma(\gamma)$, where at least two high energetic photons are required with strict cuts on activity in the tracking system. The cross-section measurements at the highest LEP energies are compared to the theory in Fig. 3.6. The angular distribution is used to test possible deviations from QED, also shown in Fig. 3.6. The lowest order expression

$$\frac{d\sigma}{d\cos\theta} = \frac{\alpha_{\text{QED}}^2 2\pi}{s}\frac{1+\cos^2\theta}{1-\cos^2\theta}, \tag{3.33}$$

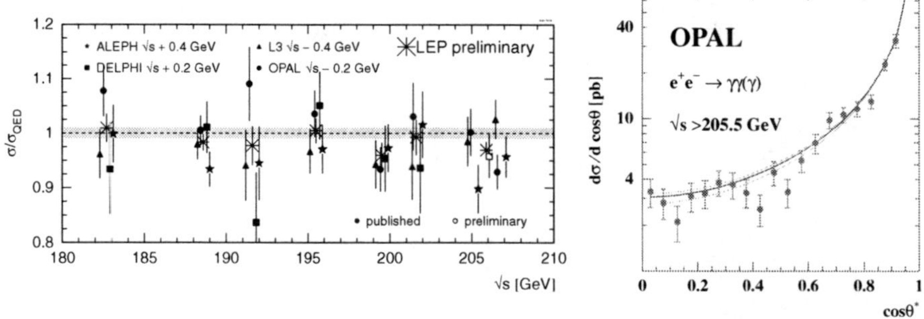

Fig. 3.6 Total cross-section of multi-photon production measured at LEP [8] and the corresponding angular distribution of photons [9] measured by OPAL, compared to QED predictions, shown as a *continuous line*. In the total cross-section calculation theoretical uncertainties are indicated as a *shaded area*. Possible deviations of the angular distribution due to exponential extensions of the Coulomb potential are shown as *dashed lines*, for $\Lambda_\pm$ values very close to the LEP combined limits

is extended by a term

$$\ldots + \frac{\alpha^2_{\mathrm{QED}} \pi s}{\Lambda_\pm}(1 + \cos^2\theta), \tag{3.34}$$

which corresponds to a short-range exponential deviation from the Coulomb potential with a cut-off parameter $\Lambda_\pm$. The LEP combined limits for this parameter are $\Lambda_+ > 392$ and $\Lambda_- > 364$ GeV at 95% confidence level [8], verifying the QED prediction nearly up to the TeV range.

3.2.2 Single Z and Z-Pair Production

At LEP energies above the Z resonance, the heavy neutral gauge boson is detected also in single Z and Z-pair production, $\mathrm{e^+e^-} \to Z/\gamma^*\mathrm{e^+e^-} \to \mathrm{f\bar{f}e^+e^-}$ and $\mathrm{e^+e^-} \to ZZ \to$ ffff, respectively. They can be imagined as the extension of QED Compton scattering and photon pair-production to virtual, off-shell photons, γ^*, whose mass eventually reaches the Z boson mass. The final state objects are therefore not the massless photons but the decay fermions of the Z.

On-shell ZZ production is described by two neutral current Feynman graphs with t- and u-channel electron exchange, the so-called NC02 set. In the fully hadronic channel ZZ $\to$ qqqq, likelihood-based analyses are applied to separate signal from the most important $\mathrm{e^+e^-} \to \mathrm{q\bar{q}}(\gamma)$ and $\mathrm{e^+e^-} \to$ WW backgrounds. Variables exploiting event shape and kinematics are combined, like, for example, the event sphericity, jet energy differences, inter-jet opening angles, jet resolution parameters, and the reconstructed Z boson masses. W-pairs are efficiently rejected by b-jet tagging. Figure 3.7 gives an example for the discrimination between ZZ signal and background on the basis of the b-quark content and the final probability variable.

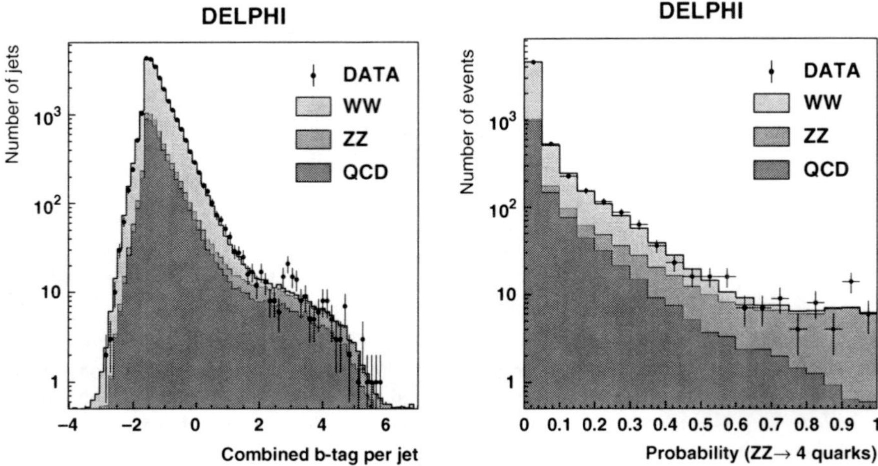

Fig. 3.7 Jet b-tagging is used to reduce W-pair background in ZZ events (*left*). This variable contributes to the WW/ZZ separation power in the combined ZZ probability for the DELPHI signal candidate events (*right*). The cross-section is derived from a fit to the probability distribution, retaining those events with a probability of at least 0.25. The distributions are taken from [10]

In the decay channel with two neutrinos, $q\overline{q}\nu\nu$ and $\ell^+\ell^-\nu\nu$, jet and lepton pairs are selected without any other activity in the detectors. Only electron and muon pairs are accepted. The visible $q\overline{q}$ and $\ell^+\ell^-$ mass as well as the recoil mass are required to be compatible with M_Z. To reject $q\overline{q}(\gamma)$ events with hard ISR photons, the missing momentum should not point along the beam direction.

In the semi-leptonic channel, $q\overline{q}\ell^+\ell^-$, with $\ell = e, \mu\ \tau$, similar kinematic criteria are imposed as in the qqqq final state, for example: compatibility with the Z boson masses, transverse momentum balance, large effective centre-of-mass energy.

The channel with the least number of ZZ candidates is the fully leptonic decay $\ell^+\ell^-\ell^+\ell^-$. The sample is however very clean with the practically only background from non-resonant $e^+e^-\ell^+\ell^-$ production. Z mass constraints are imposed either on both lepton pairs, or on the better reconstructed one and its recoil system.

In general, the cross-section is derived from a fit to the final selection variable, usually a likelihood or neural network distribution. The ZZ event rate at the highest centre-of-mass energy of 207 GeV yields, e.g., 358 candidates selected by L3 [11] in a data sample of 138.9 pb^{-1}. The signal expectation is 80.4 ± 0.1 events and the background 278.4 ± 0.6 events. The relative contributions to the ZZ signal from qqqq is about 58%, 24% from $q\overline{q}\nu\nu$, 15% from $q\overline{q}\ell^+\ell^-$, 2% from $\ell^+\ell^-\nu\nu$, and only 1% from $\ell^+\ell^-\ell^+\ell^-$.

The LEP combined ZZ cross-section [8] is illustrated in Fig. 3.8. It agrees very well with the NC02 calculation of the ZZTO [12] and YFSZZ [13] programs, which have a 2% theoretical uncertainty. The ratio of all measurements at the different centre-of-mass energies to the expectation is concentrated into a single number, R_{ZZ}, taking correlations into account, which yields [8]:

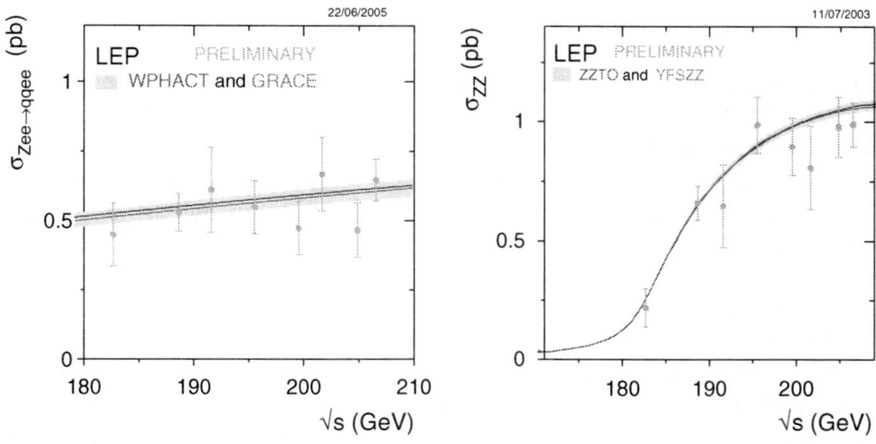

Fig. 3.8 Measured cross-sections of single Z and Z-pair production [8] at LEP, compared to theoretical predictions. The *grey band* indicates the systematic uncertainty on the theory calculation. Very good agreement is observed

$$R_{\mathrm{ZZ}}^{\mathrm{ZZTO}} = 0.952 \pm 0.052$$
$$R_{\mathrm{ZZ}}^{\mathrm{YFSZZ}} = 0.945 \pm 0.052 \,.$$

Within the 5% measurement precision, there is good agreement with the theoretical predictions. The uncertainty is dominated by statistics, and the main systematic uncertainties are from the understanding of the b-tagging, the lepton and jet energy scale, jet rates, and background cross-sections in the ZZ signal region.

Single Z production is measured at LEP in $\mathrm{e^+e^-q\bar{q}}$ and $\mathrm{e^+e^-}\mu^+\mu^-$ final states. The signal cross-section is defined as the Compton-like four-fermion process $\mathrm{e^+e^-} \rightarrow \mathrm{e^+e^-f\bar{f}}$ with the following criteria on the phase space: $M_{\mathrm{f\bar{f}}} > 60$ GeV, $\theta_{e^+} < 12^\circ$, $12^\circ < \theta_{e^-} < 120^\circ$ and $E_{e^-} > 3$ GeV. The positron is assumed to have emitted the scattering quasi-real photon, and the electron is interacting with this photon and is scattered into the detector acceptance region. The corresponding criteria apply similarly to the charged conjugate reaction.

The detector signal is thus a pair of jets or muons, compatible with an on-shell Z boson, and a single scattered electron. In addition, large missing momentum pointing along the beam direction is required. The main backgrounds in the hadronic channel are from single W's, $\mathrm{e^+e^-} \rightarrow \mathrm{qqe}\nu$, quark pair production and W pairs. In the $\mathrm{e^+e^-}\mu^+\mu^-$ channel, mainly muon-pair and two-photon production, $\mathrm{e^+e^-}\mu^+\mu^-$, need to be rejected. Figure 3.8 shows the measured cross-section as a function of centre-of-mass energy. Like for ZZ production, the measurement uncertainty is dominated by statistics, and the main systematic uncertainties are from lepton and jet energy measurement, selection efficiencies, background and signal modelling. Very good agreement with the theoretical calculations using the four-fermion programs WPHACT [14] and GRC4f [15] is obtained.

3.2.3 Z+γ Production

The simultaneous production of the neutral gauge bosons, photon and Z, is investigated at LEP in the decay channels $q\bar{q}\gamma$ and $\nu\nu\gamma$ which are the dominant Z+γ decays. In both final states, events containing an energetic photon with a recoil mass compatible with the Z boson mass are selected. In the hadronic channel, the mass of the jet-jet system must as well be close to M_Z and there should be little energy imbalance in the event. Figure 3.9 shows an example of the L3 measurement, with a distribution of the photon recoil mass in $\nu\nu\gamma$ events, and the $Z\gamma$ cross-section as a function of centre-of-mass energy.

The $Z\gamma$ process with an initial state photon, γ_{ISR}, escaping invisibly along the beam pipe is of special interest as a calibration method of the LEP beam energy. In those events, the detector signature is a jet or lepton pair, from $e^+e^- \rightarrow Z + \gamma_{ISR} \rightarrow q\bar{q} + \gamma_{ISR}$ or $e^+e^- \rightarrow Z + \gamma_{ISR} \rightarrow \ell^+\ell^- + \gamma_{ISR}$, with large missing momentum along the beam direction. The Z mass can be determined precisely from the fermion directions, according to Eq. (3.29), if $\sqrt{s'}$ is identified with the mass of the produced Z boson. The so-called radiative return to the Z is seen nicely in Fig. 3.10. The Z mass determined from the radiative events, $M_Z^{f\bar{f}}$, can be compared to the precision Z mass, M_Z, from the Z peak. Equivalently, this comparison can be translated into a test of the LEP centre-of-mass energy using:

$$\Delta\sqrt{s} = \sqrt{s} - \sqrt{s}_{\text{LEP}} = \sqrt{s}\frac{M_Z^{f\bar{f}} - M_Z}{M_Z}, \qquad (3.35)$$

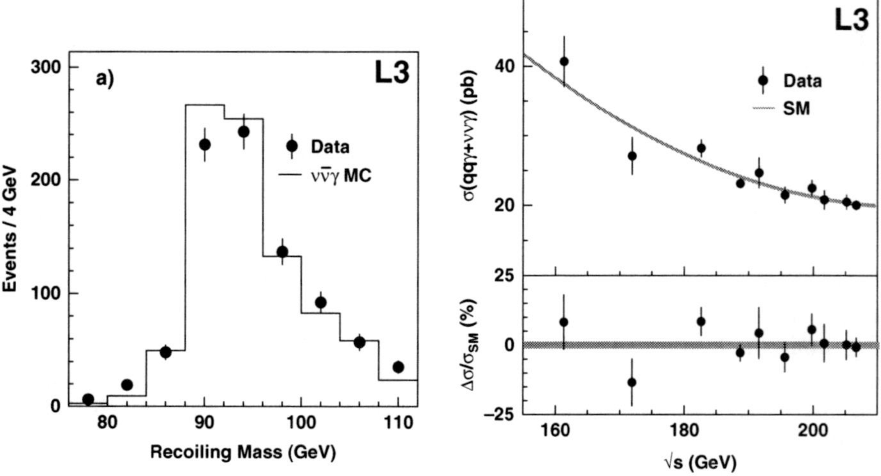

Fig. 3.9 Mass of the system recoiling to the photon in $\nu\nu\gamma$ events, as measured by L3 [16] (*left*), and the combined $Z + \gamma$ cross-section (*right*) compared to the theoretical prediction

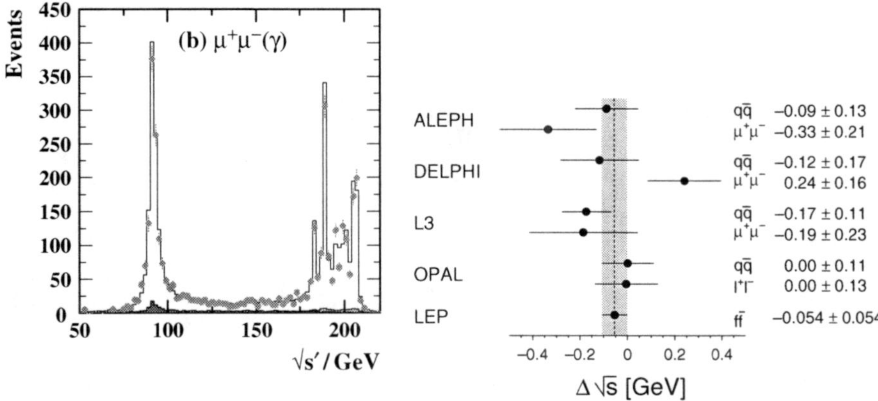

Fig. 3.10 Z mass spectrum calculated from the muon production angles, as measured by OPAL [17], for all LEP energies combined (*left*). Difference in centre-of-mass energy $\Delta\sqrt{s}$ measured by the four LEP experiments in hadronic and leptonic Z return events. The combined LEP value [18] is in good agreement with the more precise standard LEP energy calibration

with the nominal value of $\sqrt{s}_{\text{LEP}}$ [19] provided by the LEP energy working group. The LEP combined value is [18]

$$\Delta\sqrt{s} = 0.054 \pm 0.054 \ \text{MeV} \tag{3.36}$$

in good agreement with no shift with respect to the more precise standard LEP energy calibration. This is a nice confirmation of the calibration procedure and an interesting cross-check for the precise determination of M_{W}.

3.2.4 Anomalous Neutral Gauge Boson Couplings

From the measurement of ZZ and Zγ production, one can infer on the coupling structure of the neutral triple gauge boson vertex. The most general ZZV vertex, with V = Z,γ, for on-shell Z's which respects Bose symmetry can be written as [20–22]:

$$\Gamma^{\alpha\beta\mu}_{ZZV}(q_1, q_2, P) = \frac{s - m_V^2}{M_Z^2} \left\{ i f_4^{ZZV} \left(P^\alpha g^{\mu\beta} + i f_5^{ZZV} \varepsilon^{\mu\alpha\beta\rho}(q_1 - q_2)_\rho \right\}, \tag{3.37}$$

where P is the four-momentum of the incoming V boson, while q_1 and q_2 are the four-momenta of the produced Z boson pair. The f_5 coupling is CP violating while f_4 is CP conserving. In the Standard Model, both are zero at tree level.

Similarly, the ZγV vertex is described by

$$\begin{aligned}\Gamma^{\alpha\beta\mu}_{ZZV}(q_1,q_2,P) = \frac{s-m_V^2}{M_Z^2}\Big\{& h_1^V\left(q_2^\mu g^{\alpha\beta} - q_2^\alpha g^{\mu\beta}\right) \\ &+\frac{h_2^V}{M_Z^2}P^\alpha\left(P\cdot q_2 g^{\mu\beta} - q_2^\mu P^\beta\right) \\ &+h_3^V \varepsilon^{\mu\alpha\beta\rho} q_{2,\rho} \\ &+h_4^V p^\alpha \varepsilon^{\mu\beta\rho\sigma} P_\rho q_{2,\sigma}\Big\},\end{aligned} \tag{3.38}$$

where V = Z,γ is again the incoming virtual boson in the s-channel. Terms proportional to P^μ and q_1^α are omitted because they do not contribute in e^+e^- annihilation. The couplings h_1^V and h_2^V are even under parity, and h_3^V and h_4^V are CP even. All couplings are C-odd, h_1^V and h_2^V are therefore CP violating. Because of gauge invariance for V = γ both anomalous contributions vanish for $s = M_V^2$. In the terminology of an effective Lagrangian, the interaction is induced by operators of dimension six and higher.

The h_i^V couplings are determined from the kinematic observables in Z+γ events, like the photon energies and angles and, if measured, the fermion energies and angles. Similarly, the full ZZ event kinematics is exploited to determine the f_i^V couplings. The results of all LEP data combined [8] are shown in Figs. 3.11 and 3.12. All couplings are found to be consistent with zero. The gauge boson coupling structure in the neutral sector fully corresponds to the Standard Model expectations.

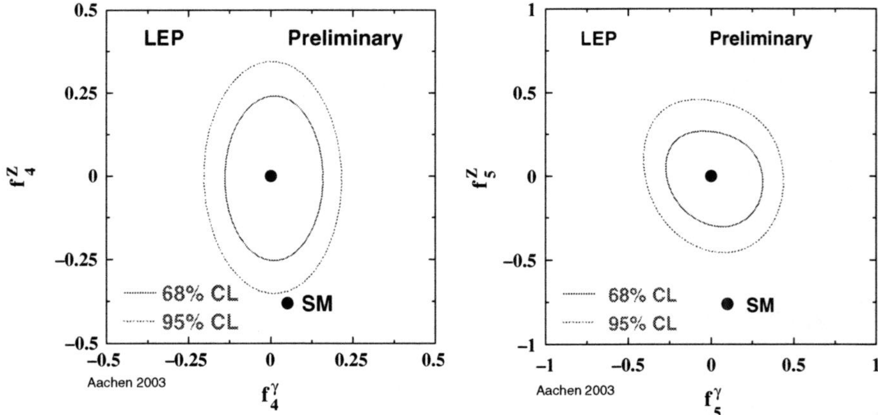

Fig. 3.11 Contour curves of 68 and 95% confidence level limits on the neutral triple gauge couplings f_4^V and f_5^V [8]

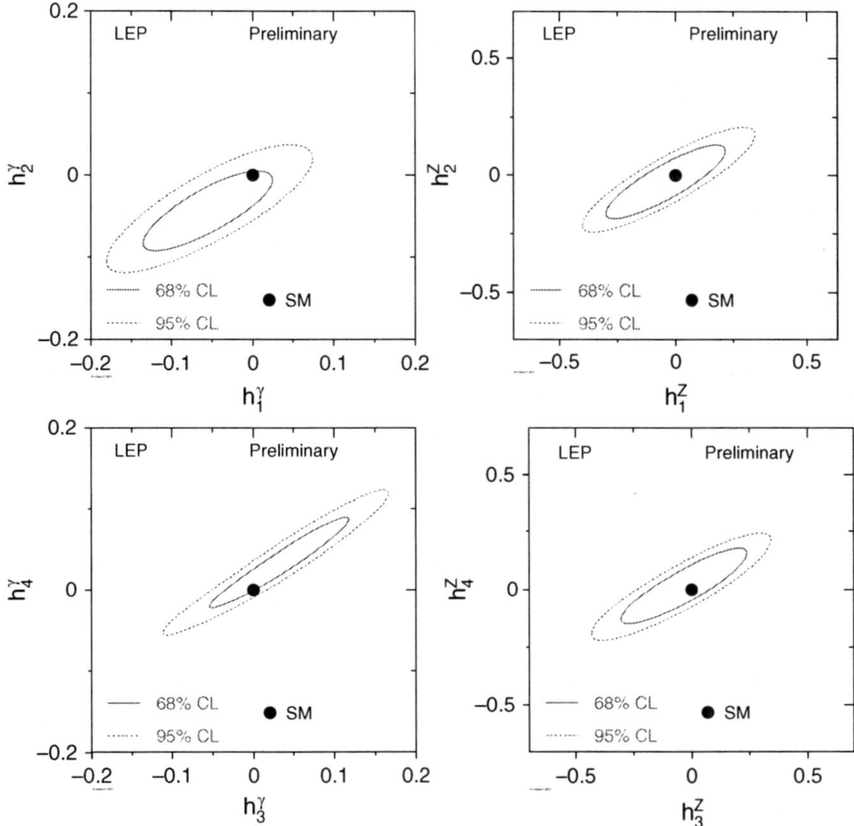

Fig. 3.12 Contour curves of 68 and 95% confidence level limits on the neutral triple gauge couplings h_i^V [8]

3.3 Measurement of the W-Pair Cross-Section

The three main final state categories for W-pair production at LEP are fully leptonic, $\ell\nu\ell\nu$, semi-leptonic, $qq\ell\nu(\gamma)$, and fully hadronic, $qqqq(\gamma)$, with branching fractions of 43.5, 46.2 and 10.3%, respectively. All events may be accompanied by photon radiation, indicated by (γ), either in the initial or final state. The events are detected by identifying the visible W decay products, the leptons and quark jets. All lepton flavours, including leptonic and hadronic tau decays, are considered in the selection channels.

The LEP experiments apply different strategies to retain the signal events in data and to reject the different backgrounds. Cut-based and likelihood-based selections are applied, as well as neural network techniques [23–26]. They all exploit the kinematic properties of the events.

The main selection criteria for $\ell\nu\ell\nu$ events are two charged leptons and missing energy. The leptons are required to be acoplanar to reduce the dominating backgrounds from lepton-pair production and possible cosmic rays. The latter is further suppressed by requiring an event timing compatible with the beam crossing. A veto on other activity in the detector is furthermore applied. The main remaining background is from two-photon collision processes. Figure 3.13 shows an example for the measured acoplanarity distribution in L3 data. The purity of the event samples are around 70–80% and the selection efficiencies between 20 and 30% in the hadronic $\tau\nu\tau\nu$ channel up to 60% for $e\nu e\nu$ and $e\nu\mu\nu$ events. The cross-efficiencies, especially for events with $W \to \tau\nu_\tau$ decays, can be up to 10%.

Semi-leptonic $qq\ell\nu$ events are selected by their hadronic activity, two jets and one energetic and isolated lepton in the final state, and high invariant masses of the jet-jet system. The neutrino momentum is actually well reconstructed in $qqe\nu$ and $qq\mu\nu$ events by identification with the missing momentum. A large $\ell\nu$ mass is used as selection criteria as well. The missing momentum is required to not point along the beam direction. An example for a variable combining lepton-jet separation and missing momentum direction is shown in Fig. 3.13. In the $qq\tau\nu$ channel with leptonic tau decays, a low $\ell\nu$ mass is used to separate this decay from the $qqe\nu$ and $qq\mu\nu$ events. This minimises the overlap between the semi-leptonic channels, which is at the 5–10% level. Hence, it reduces correlations in the measurement of the W branching fractions. The main background in the semi-leptonic channels is from $e^+e^- \to q\bar{q}(\gamma)$ and Z-boson pair production. Efficiencies are between 50 and 65% in the tau channel and 75–90% in the electron and muon channel. The experiments with a larger angular coverage for efficient lepton detection generally obtain higher

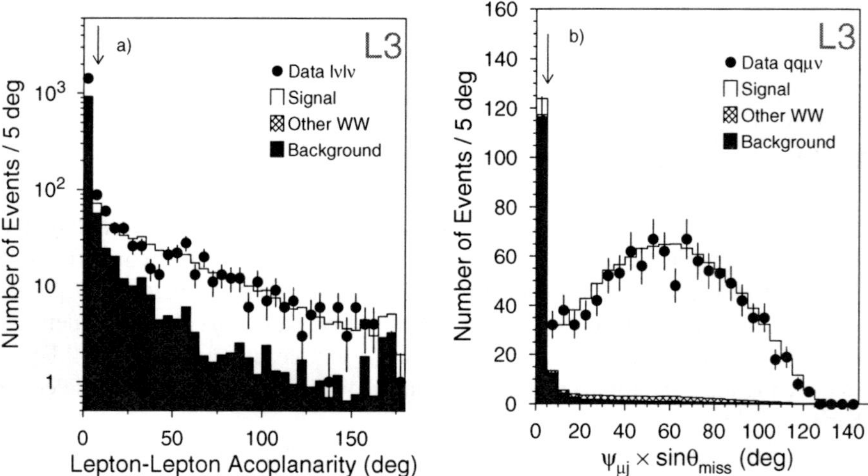

Fig. 3.13 Example of W-pair selection variables used by the L3 experiment [25]: (**a**) acoplanarity of $\ell\nu\ell\nu$ candidates, (**b**) the angle between the muon and the next jet in $qq\mu\nu$ candidate events, multiplied by the sine of the missing momentum

efficiencies. The purity is as high as 98% and above for qqeν and qq$\mu\nu$ events and reaches 65–85% in the qq$\tau\nu$ channel.

The selection of fully hadronic W-pair decays is based on little missing energy, high multiplicity and four-jet topology. Jet resolution parameters, like the Durham y_{34} variable [27], discriminate against quark-pair production with additional gluon radiation, $e^+e^- \to q\bar{q}g$. By applying a neural network based selection which uses different kinematic and topological quantities the main background from $e^+e^- \to q\bar{q}gg$ and hadronic Z-pairs is further reduced. At a selection efficiency of 85–90% a purity of 80% is reached. Figure 3.14 gives examples for the spherocity distribution and the neural network output measured by L3.

Each experiment retains 700–1,000 $\ell\nu\ell\nu$ candidates, 1,200–1,500 qq$\ell\nu$ candidates, and 5,000–5,500 qqqq candidates, from which the W-pair cross-section is derived. The total production cross-section for W-pairs is defined as the corresponding CC03 process, which means that other diagrams that lead to the same final state are considered as background. The efficiencies that enter the calculation are calculated with the state-of-the art Monte Carlo programs that include full $O(\alpha)$ electroweak corrections [28], like the KandY [29], WPHACT [14] and RacoonWW [30] event generators. Cross-checks between the different calculations are performed and yield consistent results.

The W-pair cross-sections are determined in each decay channel separately and cross-efficiencies for other channels are taken into account. Systematic uncertainties arise mainly from the modelling of the detector response to the measured leptons and quark jets, which is between 1 and 2% of the cross-section. Smaller contributions are from hadronisation modelling, where the string-fragmentation model Pythia, and the colour-dipole models Herwig and Ariadne are compared. These models are tuned on Z decay data depleted in $b\bar{b}$ final states, which are the most

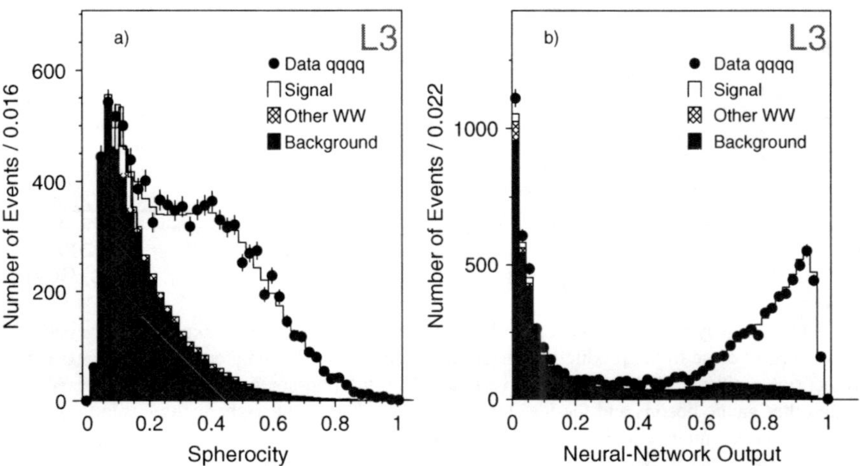

Fig. 3.14 (**a**) The spherocity in fully hadronic events is one of the inputs to the neural network analysis of the L3 experiment [25]. (**b**) The final neural network output of the qqqq selection [25]

similar to the hadronic W decays. Furthermore, there are uncertainties due to Final State Interactions (see Sect. 3.6), photon radiation and the variation of the W mass and width within their measurement uncertainties. Luminosity uncertainties are considered as well. Altogether, the systematic uncertainties are in the order of 1.5–2.5%, much smaller than the statistical uncertainty for each channel and energy point.

The total W-pair cross-section is determined by combining all decay channels at each energy points and assuming the W branching fractions to follow the Standard Model prediction. When combining data of different experiments, correlations between systematic uncertainties are properly considered. Already with the data close to the WW threshold, there is clear evidence for the contribution of all three lowest order diagrams to the W-pair production, as can be seen in Fig. 3.15.

Each single contribution from t-channel neutrino exchange, or s-channel γ or Z boson exchange, shown in Fig. 1.8, would lead to a steadily increasing cross-section with centre-of-mass energy and would eventually violate unitarity. Only the coupling of the W boson to the other gauge bosons, γ and Z, caused by the non-abelian nature of the $SU(2)$ gauge group, and the interference between the three contributions guarantees correct high-energy behaviour and agrees with the measured W-pair cross-section.

The result of the combined LEP measurement for $\sqrt{s} = 161 - 209$ GeV is shown in Fig. 3.15. The accuracy of the theoretical prediction increases from 0.7% at $\sqrt{s} = 170$ to 0.4% at $\sqrt{s}$ above 200 GeV. At the threshold, where the W bosons are practically on-shell, the latest LPA/DPA techniques can not be applied and the

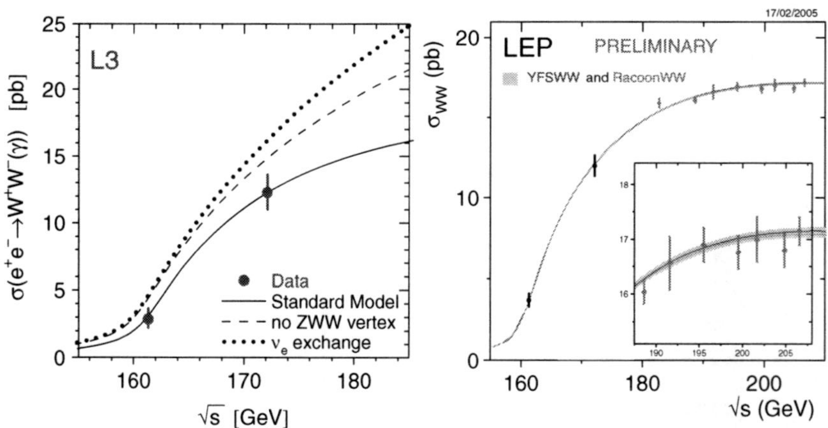

Fig. 3.15 (**a**) Early measurements of the W-pair cross-section by the L3 experiment for centre-of-mass energies close to the production threshold, compared to the Standard Model calculation and predictions omitting WWZ and WWγ vertices. Only the non-abelian couplings of the W boson to the other gauge bosons and the interference between the t- and s-channel contributions avoids the violation of unitarity and agrees with the measured W-pair cross-section. (**b**) LEP combined measurement of the W-pair cross-section as a function of the centre-of-mass energy compared to latest theoretical predictions [8]

uncertainty on the prediction reaches 2%. All LEP data are in excellent agreement with the Standard Model prediction over the whole energy range.

The agreement between the theoretical prediction is summarised in a single R parameter, which is the combined ratio of measurement to theory for the whole energy range:

$$R_{\mathrm{WW}} = \frac{\sigma_{\mathrm{WW}}^{\mathrm{meas}}}{\sigma_{\mathrm{WW}}^{\mathrm{theo}}} \tag{3.39}$$

The extracted R values are [8]:

$$R_{\mathrm{WW}}^{\mathrm{YFSWW}} = 0.994 \pm 0.009 \tag{3.40}$$

$$R_{\mathrm{WW}}^{\mathrm{RACOONWW}} = 0.996 \pm 0.009 \tag{3.41}$$

when comparing with the YFSWW and RacoonWW predictions. A very good agreement is observed within the experimental precision of 1%, which is close to the theoretical uncertainty of 0.5% [28].

From the measurements of the cross-sections in the individual decay channels the W branching fractions are determined, where the total sum is assumed to be unity. The results of the LEP experiments are shown in Fig. 3.16. The Standard Model expectation of the hadronic branching fraction, $Br(\mathrm{W} \to \mathrm{q\overline{q}})$, is 67.51%.

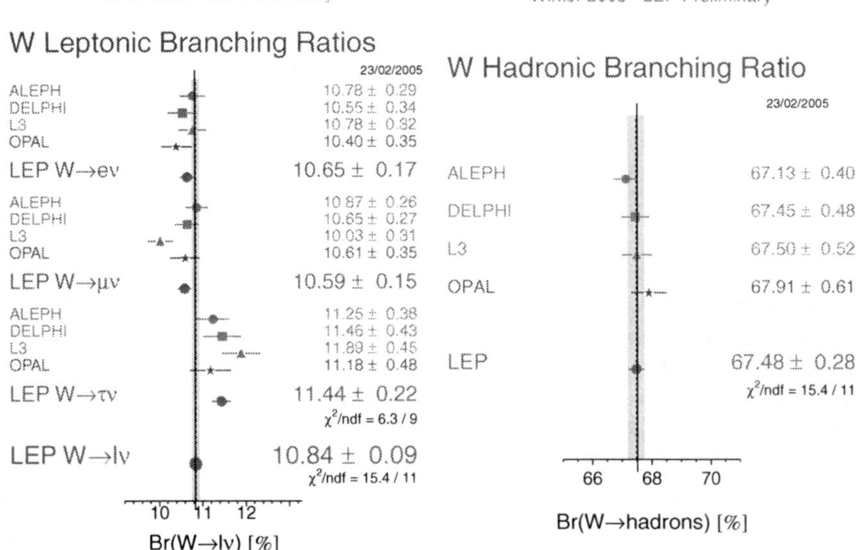

Fig. 3.16 Branching fractions of the W boson as measured by the LEP experiments and their combination [8]

The leptonic ones are expected to be all equal with $Br(\mathrm{W} \rightarrow \ell\nu) = 10.83\%$. As it is evident from the results of all LEP experiments, there is an excess of the branching ratio $W \rightarrow \tau\nu$, which is at a level of 2.8 standard deviations:

$$\frac{2Br(\mathrm{W} \rightarrow \tau\nu)}{Br(\mathrm{W} \rightarrow \mathrm{e}\nu)Br(\mathrm{W} \rightarrow \mu\nu)} = 1.077 \pm 0.026. \tag{3.42}$$

Assuming lepton universality, the combined results of:

$$Br(\mathrm{W} \rightarrow \ell\nu) = 10.84 \pm 0.06(\mathrm{stat.}) \pm 0.07(\mathrm{syst.}) \tag{3.43}$$

$$Br(\mathrm{W} \rightarrow \mathrm{q\bar{q}}) = 67.48 \pm 0.18(\mathrm{stat.}) \pm 0.21(\mathrm{syst.}) \tag{3.44}$$

are however in good agreement with the expectations [8].

3.4 Measurement of Single-W Production

The production of single W bosons, $\mathrm{e^+e^-} \rightarrow We\nu$, is defined as the complete t-channel subset of Feynman diagrams leading to the $e\nu\mathrm{ff}$ final state [8]. To suppress multi-peripheral graphs the process is further characterised by the following phase space requirements on the $\mathrm{q\bar{q}}$ mass, the lepton energy and the electron and positron scattering angles:

$$\begin{aligned} m_{\mathrm{q\bar{q}}} > 45\ \mathrm{GeV} &\quad \text{for} \quad We\nu \rightarrow \mathrm{e}\nu\mathrm{qq}\ , \\ E_\ell > 20\ \mathrm{GeV} &\quad \text{for} \quad We\nu \rightarrow \mathrm{e}\nu\tau\nu \\ &\quad \text{and} \quad We\nu \rightarrow \mathrm{e}\nu\mu\nu\ , \\ E_{e^+} > 20\ \mathrm{GeV},\ |\cos\theta_{e^+}| < 0.95,\ |\cos\theta_{e^-}| > 0.95 &\quad \text{for} \quad We\nu \rightarrow \mathrm{e}\nu\mathrm{e}\nu\ . \\ \text{(and the charge conjugate)} & \end{aligned}$$

The main signature of single-W production is the forward scattered electron, which remains invisible in the detector, and the decay products of the W that can be measured [31–33]. In L3, the hadronic W decay is therefore selected by requiring a visible energy in the calorimeters compatible with a W decay, $0.30 < E_{\mathrm{vis}}/\sqrt{s} < 0.65$. The missing momentum vector should not point along the beam direction, $|\cos\theta_{\mathrm{miss}}| < 0.92$, because this is the typical signature of radiative $\mathrm{e^+e^-} \rightarrow \mathrm{q\bar{q}}(\gamma)$ events with ISR photons escaping along the beam pipe. Events with three-jet topology from $\mathrm{qq}\tau\nu$ W-pair production are also removed. Kinematic criteria like the invariant jet-jet mass and velocity are combined in a likelihood variable to suppress $\mathrm{e^+e^-} \rightarrow \mathrm{ZZ} \rightarrow \nu\nu\mathrm{q\bar{q}}$ events. Eventually, a neural network based on visible energy, visible mass, single-W kinematics and 2/3-jet topology is applied to select the final event sample.

Leptonic single-W final states have the striking signature of single leptons in the detector without any other visible activity. The recoil mass to the lepton must be compatible with single-W production, which also rejects $e^+e^- \rightarrow \nu\nu\gamma$ events with converted photons.

Figure 3.17 shows two selection criteria for hadronic and leptonic W decays. The hadronic W final states are triggered redundantly, while trigger efficiencies for leptonic final states are about 90% for muons, 95% for electrons and close to 100% for taus. They are determined directly from data.

The measured cross-sections combined for the LEP experiments is shown in Fig. 3.18 for the hadronic W decays and all channels combined. The theoretical predictions are less precise than for W-pair production and reach a precision of 5%. This is estimated from two different calculations in fixed-width and fermion-loop scheme [34], which consists in including all fermionic one-loop corrections in tree-level amplitudes and resumming the self-energies. Data and theory agree well over the whole LEP energy range. The global agreement is expressed in the measurement-to-theory ratio R as [8]:

$$R_{We\nu}^{\mathrm{GRC4f}} = 1.051 \pm 0.075 \tag{3.45}$$

$$R_{We\nu}^{\mathrm{WPHACT}} = 1.083 \pm 0.078 \tag{3.46}$$

for the two theoretical programs GRC4f [15] and WPHACT [14]. The measurements show again that the W coupling structure is in agreement with the Standard Model, especially the WWγ vertex contributions. This is further investigated in the next chapter.

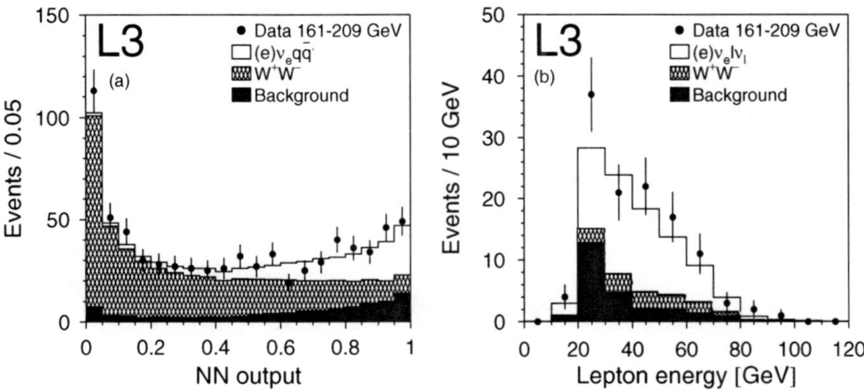

Fig. 3.17 (**a**) Neural network output as final selection variable for hadronic single-W events in L3 [33]. (**b**) Lepton energy spectrum in single-W decays as measured by L3 [33]

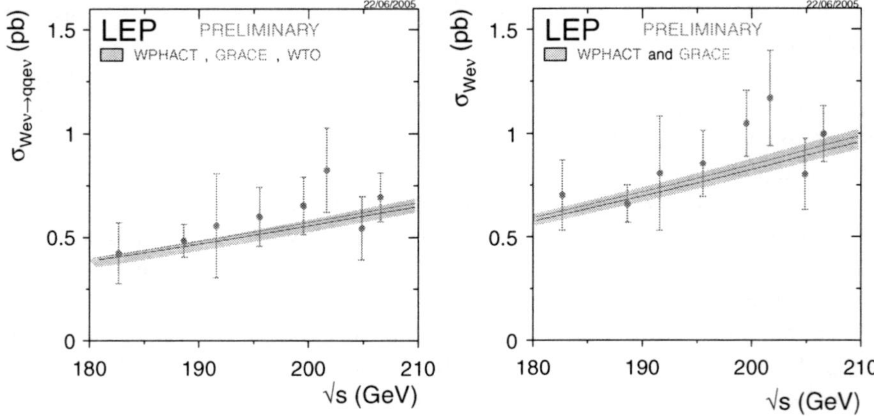

Fig. 3.18 (**a**) Measured cross-section of hadronic single-W production at LEP as function of centre-of-mass energy. The data are compared to the theoretical predictions of WPHACT [14], GRC4f [15] and WTO [34]. (**b**) Total single-W cross-section for the same energy range, compared to WPHACT [14] and GRC4f [15]

3.5 Determination of Triple Gauge Boson Couplings

In the Standard Model, W bosons couple to the other gauge bosons, γ and Z, by means of their charge and the weak coupling constant, respectively. To test the coupling structure it is helpful to extend the Standard Model to a general WWV vertex, with V = γ,Z. In this way, additional coupling constants are introduced which describe possible low energy manifestations of new physics beyond the Standard Model. Both, the Standard Model and anomalous couplings can then be measured or constrained by data.

The WWV vertex is generally parameterised in a phenomenological effective Lagrangian [20, 21]:

$$\begin{aligned} i\mathcal{L}_{eff}^{WWV} = g_{WWV} \Big[& g_1^V V^\mu \left(W_{\mu\nu}^- W^{+\nu} - W_{\mu\nu}^+ W^{-\nu}\right) + \kappa_V W_\mu^+ W_\nu^- V^{\mu\nu} \\ & + \frac{\lambda_V}{M_{\mathrm{W}}^2} V^{\mu\nu} W_\nu^{+\rho} W_{\mu\rho}^- + i g_5^V \varepsilon_{\mu\nu\rho\sigma} \left\{ \left(\partial^\rho W^{-\mu}\right) W^{+\nu} - W^{-\mu} \left(\partial^\rho W^{+\nu}\right) \right\} V^\sigma \\ & + i g_4^V W_\mu^- W_\nu^+ \left(\partial^\mu V^\nu + \partial^\nu V^\mu\right) \\ & - \frac{\tilde{\kappa}_V}{2} W_\mu^- W_\nu^+ \varepsilon^{\mu\nu\rho\sigma} V_{\rho\sigma} - \frac{\tilde{\lambda}_V}{2M_{\mathrm{W}}^2} W_{\rho\mu}^- W_\nu^{+\mu} \varepsilon^{\nu\rho\alpha\beta} V_{\alpha\beta} \Big] \end{aligned}$$

The overall couplings are defined as $g_{WW\gamma} = e$ and $g_{WWZ} = e \cot\theta_{\mathrm{w}}$, which are the W electromagnetic charge and weak coupling to the Z. The W and V field strengths are here defined as: $W_{\mu\nu} = \partial_\mu W_\nu - \partial_\nu W_\mu$ and $V_{\mu\nu} = \partial_\mu V_\nu - \partial_\nu V_\mu$. For real photons ($Q^2 = 0$), g_1^γ and g_5^γ are fixed by gauge invariance to 1 and 0, respectively. In the Standard Model the only non-zero couplings are

$$g_1^Z = g_1^\gamma = \kappa_Z = \kappa_\gamma = 1 \tag{3.47}$$

at tree level. The terms proportional to g_1^V, κ_V and λ_V are C and P conserving, while g_5^V violates C and P but conserves CP. Violation of CP is parameterised by g_4^V, $\tilde{\kappa}_V$ and $\tilde{\lambda}_V$.

The lowest order terms of the multipole expansion of the W-photon interaction are directly related to the couplings. The charge, Q_W, the magnetic dipole moment, μ_W, and the electric quadrupole moment, q_W, of the W^+ are given by:

$$Q_W = eg_1^\gamma \tag{3.48}$$

$$\mu_W = \frac{e}{2M_\mathrm{W}}\left(g_1^\gamma + \kappa_\gamma + \lambda_\gamma\right) \tag{3.49}$$

$$q_W = -\frac{e}{M_\mathrm{W}^2}\left(\kappa_\gamma - \lambda_\gamma\right) \tag{3.50}$$

One can obtain theoretical constraints on the triple-gauge boson couplings (TGC) by asking for a global "custodial" $SU(2)$ symmetry [35] of the effective Lagrangian. This is supported by experimental data since it avoids deviations of the ρ parameter from the well established value close to 1. The additional symmetry leads to the following relations between the C and P violating couplings:

$$\kappa_Z = g_1^Z - \tan^2\theta_\mathrm{w}(\kappa_\gamma - 1) \tag{3.51}$$

$$\lambda_Z = \lambda_\gamma \tag{3.52}$$

With this assumption the parameter space becomes more restricted and constraints are more stringent with the given amount of data.

The charged TGCs are extracted from LEP data in an analysis of the multi-differential cross-section of W-pair production [36–39]. Neglecting photon radiation and fixing the mass of the W boson, five angles describe the four-fermion final state in W-pair decays: the polar decay angle of the W^- boson, Θ_W, and the polar and azimuthal decay angles of the fermions in the W rest frames. The TGCs influence the total production cross-section, the W^- production angle and the fermion angles by changing the W polarisation.

To improve the resolution on the angles, a kinematic fit is applied to the events, asking for four-momentum conservation and equality of the two reconstructed W masses. In $\mathrm{qq}\tau\nu$ events the two hadronic jet energies are rescaled such that their sum equals $\sqrt{s}/2$. This yields nearly the same performance as a kinematic fit, which can be less constrained due to the tau neutrino. In fully hadronic events, the assignment of two jet pairs to two W decays, the so-called "pairing", can be done using neural network techniques. They yield a correct pairing typically in the order of 75–80% of the cases [38]. Input to the neural network are kinematic variables like the difference and sum of the masses of the jet pairs, and the sum and minimum of the angles between paired jets. Another quantity is the value of the matrix element for the reaction $\mathrm{e^+e^-} \to \mathrm{WW} \to \mathrm{ffff}$ applied to the four reconstructed jets and calculated, e.g., with the EXCALIBUR [40] program. Finally, the difference between the W charges is used as calculated from the sum of jet charges.

The jet-charge is assigned by applying a track based technique. The tracks belonging to each quark jet are weighted by their momenta:

$$q_{\text{jet}} = \frac{1}{N_{\text{tracks}}} \sum_{\text{tracks}} q_i |\mathbf{p}_i \cdot \mathbf{d}_{\text{jet}}|^{\kappa} , \tag{3.53}$$

with the number of tracks, N_{tracks}, their charge, $q_i = \pm 1$, and the track momentum $\mathbf{p}_i$ projected on the jet direction $\mathbf{d}_{\text{jet}}$. The momentum weight κ is chosen to be unity. With this definition, the charge is found to be correct in about 70% [38] of the Monte Carlo events under the condition that the pairing is correct. Alternatively, if the jet charge is not used in the analysis [36] the corresponding angular distribution are folded and no sign is determined.

The W^- direction, $\cos\Theta_W$, in semi-leptonic events is determined from the hadronic W decay. The sign of $\cos\Theta_W$ is derived from the lepton charge. In fully hadronic events, the direction comes from either W pair, which are back-to-back after the kinematic fit. Sign information is provided by the charges of the jet-pairs. Figure 3.19 shows the W boson production angle in W-pairs as measured by the ALEPH experiment, and the sensitivity of the data to the anomalous λ_γ coupling.

Additional sensitivity to the WWγ vertex is also in single-W production, where the hadronic cross-section and the leptonic energy spectrum are used as input. Overlap with W-pair events is carefully removed to avoid double-counting. Single photon events, $e^+e^- \rightarrow \nu_e\bar{\nu}_e\gamma$, are mainly produced through initial-state radiation in s-channel Z-boson exchange or t-channel W-boson exchange, but the process contains also a small contribution from W-boson fusion through the WWγ vertex.

All measured final states are input to a combined analysis of the angular distributions, energy spectra and the production rates at the different centre-of-mass energies. The dependence of the observables on the couplings is derived by mainly using two different techniques, the optimal observable [41] and matrix-element reweighting methods. The former is exploiting the fact that the differential cross-section, $d\sigma/d\Omega$, in any set of measurable variables, Ω, depends only quadratically on the couplings, g_i:

$$\frac{d\sigma}{d\Omega} = S_0 + S_i g_i + S_{ij} g_i g_j . \tag{3.54}$$

It can then be shown that the set of observables $\mathcal{O}_i = S_i/S_0$ and $\mathcal{O}_{ij} = S_{ij}/S_0$ are optimal in the sense that they contain the maximum information about the couplings g_i. The coupling parameters, g_i $(i = 1, \ldots, n)$, are determined by comparing the measured mean values of $\mathcal{O}_i$ and $\mathcal{O}_{ij}$ to their expectation values, $E[\mathcal{O}_i]$ and $E[\mathcal{O}_{ij}]$, for a certain set of couplings. The optimal observables can be either calculated analytically or by Monte Carlo simulations. Correlations between them are properly taken into account.

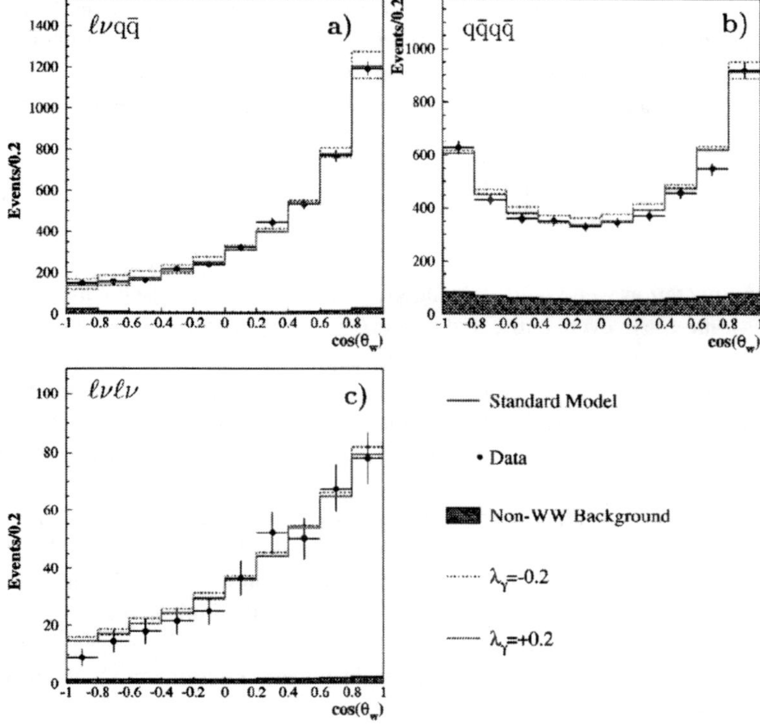

Fig. 3.19 Distribution of the W boson production angle in W-pairs as measured by the ALEPH experiment [36]. Data are combined from all LEP centre-of-mass energies. Events from fully hadronic final states enter with a weight corresponding to the probability of correct W charge measurement. Semi-leptonic events yield a weight of one, while fully leptonic final states are considered with a weight of 0.5. Data are shown together with the Standard Model prediction and the expectation for a λ_γ value of ± 0.2

The expectation values, $E[\mathcal{O}(g_1, \ldots, g_n)]$, are usually constructed by reweighting of simulated Monte Carlo events. This reweighting is based on the matrix element squared, $|M|^2$, calculated for the given production process, e.g. $e^+e^- \rightarrow$ WW $\rightarrow$ ffff(γ). A weight for each simulated signal event j is then determined as the ratio

$$w_j(g_1, \ldots, g_n) = \frac{\left|M\left(p_1^j, \ldots, p_n^j; g_1, \ldots, g_n\right)\right|^2}{\left|M_{SM}\left(p_1^j, \ldots, p_n^j\right)\right|^2}, \tag{3.55}$$

where the four-momenta of the final state particles are denoted as $p_1^j, \ldots, p_n^j$. The matrix element M contains anomalous contribution parameterised by $g_1, \ldots, g_n$, while M_{SM} is the matrix element for the Standard Model expectation.

The second extraction method is based directly on matrix element reweighting and compares the measured differential cross-section with the predicted ones. A likelihood is constructed as the Poisson probability to observe N_i events in a certain phase space interval (bin), when $\mu_i(g_1, \ldots, g_n)$ events are expected:

$$L = \prod_{i=1}^{N_{\text{bins}}} \frac{e^{-\mu_i(g_1,\ldots,g_n)} \mu_i(g_1, \ldots, g_n)^{N_i}}{N_i!}. \tag{3.56}$$

The expectation value μ_i is calculated by taking the normalised sum of the simulated Monte Carlo events, rescaled by the event weights $w_j(g_1, \ldots, g_n)$.

The ALEPH and OPAL experiments apply an optimal observable analysis [36, 39] and the L3 experiment uses the reweighting method to determine the charged TGCs [38]. DELPHI derives the couplings from the spin-density matrix [37]. All experiments exploit furthermore the dependency of the total signal cross-section sections on the couplings.

Systematic uncertainties of the TGC measurement are from various sources, like background estimation, modelling of hadronisation, LEP beam energy, lepton and jet measurement, charge confusion, and final state interaction in W-pairs, none of them actually dominating. With the more accurate theoretical predictions of the W-pair production cross-section, σ_{WW}, applying LPA/DPA techniques, the uncertainty on σ_{WW} are much reduced, as well as the uncertainties on the $O(\alpha)$ corrections to the distribution of W production angle.

The likelihood curves of the LEP experiment for the different gauge couplings are combined taking correlations between systematic uncertainties into account. The current preliminary results of 1-parameter fits to the data are [8]

$$\begin{aligned} g_1^Z &= 0.991^{+0.022}_{-0.021} \\ \kappa_\gamma &= 0.984^{+0.042}_{-0.047} \\ \lambda_\gamma &= -0.016^{+0.021}_{-0.023} \,. \end{aligned}$$

The main systematic uncertainty is from $O(\alpha)$ corrections which contributes with an uncertainty of 0.010 to g_1^Z and λ_γ and with 0.010 to κ_γ. The LEP TGC measurements are however dominated by statistical uncertainties.

All couplings agree well with the theoretical expectations, as can be seen in Fig. 3.20 for two-parameter fits. This is a direct evidence for the non-abelian structure of the theory. The W boson couples indeed to the photon and the Z boson. The measurements are fully compatible with the predicted vertex structure and coupling strength of the Standard Model. Furthermore, no evidence for neutral three-boson coupling is found, neither at LEP [8] nor in pp processes at the Tevatron [42].

Anomalous contributions to the quartic couplings of the W boson, with four-boson vertices WW$\gamma\gamma$ and WWZγ, are as well constrained by measurements [43] of the photon spectra in WWγ and $\nu\nu\gamma\gamma$ events. The combined LEP results for the WWγ cross-sections are shown in Fig. 3.21. The theoretical framework is given by the following Lagrangian containing dimension-6 operators [44]:

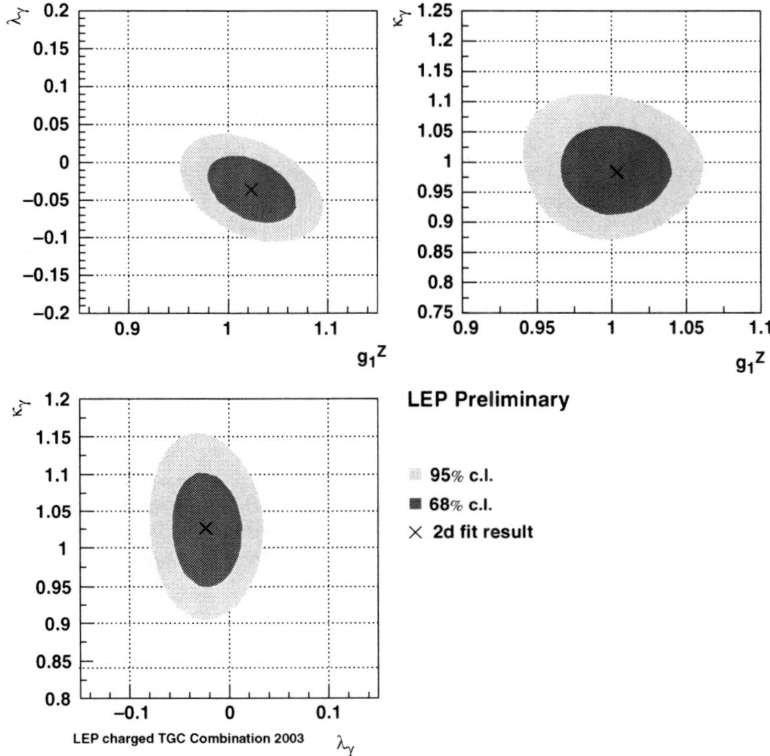

Fig. 3.20 Contour curves for the LEP measurement of the charged triple gauge boson couplings [8]

$$\mathcal{L}_6^0 = -\frac{\pi\alpha}{4}\frac{a_0}{\Lambda^2}F_{\mu\nu}F^{\mu\nu}\mathbf{W}^\alpha \cdot \mathbf{W}_\alpha \tag{3.57}$$

$$\mathcal{L}_6^c = -\frac{\pi\alpha}{4}\frac{a_c}{\Lambda^2}F_{\mu\alpha}F^{\mu\beta}\mathbf{W}^\alpha \cdot \mathbf{W}_\beta \tag{3.58}$$

$$\mathcal{L}_6^n = -\frac{\pi\alpha}{4}\frac{a_n}{\Lambda^2}\varepsilon_{ijk}W_{\mu\alpha}^{(i)}W_\nu^{(j)}W^{(k)\alpha}F^{\mu\nu}, \tag{3.59}$$

where a_0, a_c, and a_n are the anomalous quartic couplings of the W boson and Λ the energy scale of the effective theory. The influence of the anomalous couplings on the photon spectrum in WWγ events is demonstrated in Fig. 3.21. Limits are set by the LEP experiments, which are all compatible with the non-existence of anomalous contributions. They are in the order of

$$-0.020\ \text{GeV}^{-2} < a_0/\Lambda^2 < 0.020\ \text{GeV}^{-2} \tag{3.60}$$

$$-0.064\ \text{GeV}^{-2} < a_c/\Lambda^2 < 0.032\ \text{GeV}^{-2} \tag{3.61}$$

$$-0.18\ \text{GeV}^{-2} < a_n/\Lambda^2 < 0.14\ \text{GeV}^{-2}, \tag{3.62}$$

citing the DELPHI result [43] as a typical example.

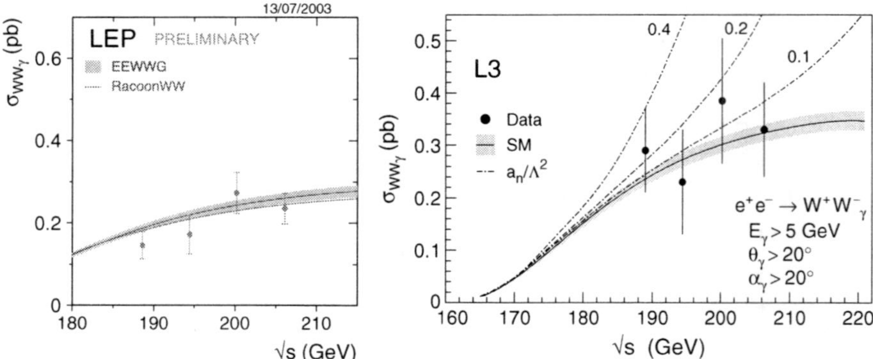

Fig. 3.21 (**a**) LEP measurement of the WWγ production [8] compared to the theoretical predictions, showing very good agreement. The signal process is defined by the following requirements on the photon kinematics: $E_\gamma > 5$ GeV, $|\cos\theta_\gamma| < 0.95$, $|\cos\theta_{\gamma f}| < 0.90$, where $\theta_{\gamma f}$ is the isolation angle between the photon and the next fermion. In addition the mass of the fermion pair from the W decay must agree with the nominal W mass, M_W, within twice the W width. (**b**) Energy spectrum of photons in WWγ events, as measured by L3. The data are compared to the Standard Model prediction and to anomalous scenarios with different effective coupling values, a_n/Λ^2=0.1, 0.2, and 0.4 [43]

The WWZZ and WWWW scattering vertices can not be directly probed at LEP. But anomalous behaviour would show up in the neutral and charged TGC measurements. The ALEPH collaboration, for example, interpreted the measurements of W-pair production and corresponding angular distributions in terms of strong $W_L W_L$ scattering with an intermediate techni-ρ exchange. Since data were found to be in agreement with Standard Model expectations a lower limit on the mass of the techni-ρ could be set at 600 GeV (95% C.L.) assuming that the width of the new particle it at most as large as its mass [36].

In summary, measurements of the electroweak boson couplings at LEP as well as at other collider experiments agree very well with the Standard Model predictions, both in their structure as well as in their magnitude.

3.6 Final State Interactions in W Boson Decays

When W boson pairs decay hadronically, so-called Bose-Einstein correlations and colour reconnection effects may alter the hadronic final state. These phenomena also influence measurements of W boson properties, especially the determination of the mass of the W.

3.6.1 Bose-Einstein Correlations

The origin of Bose-Einstein correlations (BEC) is the quantum mechanical requirement that the decay amplitude is symmetric under the exchange of identical bosons in the final state. The idea was developed in astronomy as intensity interferometry, also known as Hanbury/Brown/Twiss interferometry [45]. In particle physics, charged pion correlations were first observed in $p\bar{p}$ collisions [46–48].

In the plane-wave approximation, the amplitude to observe two identical bosons, that are produced at space-time point x_1 and x_2 with momenta p_1 and p_2 and which are measured by two detectors at points x_A and x_B is given by [47, 49]:

$$
\begin{aligned}
|p_1 p_2\rangle &= \frac{1}{\sqrt{2}} \left\{ e^{ip_1(x_1-x_A)+i\phi_1} e^{ip_2(x_2-x_B)+i\phi_2} \right. \\
&\quad \left. + e^{ip_1(x_2-x_A)+i\phi_1} e^{ip_2(x_1-x_B)+i\phi_2} \right\} \\
&= \frac{1}{\sqrt{2}} e^{i(p_1 x_A p_2 x_B - \phi_1 - \phi_2)} \left\{ e^{ip_1 x_1} e^{ip_2 x_2} + e^{ip_1 x_2} e^{ip_2 x_1} \right\},
\end{aligned}
\tag{3.63}
$$

where ϕ_1 and ϕ_2 are the phases. The last term arises due to the symmetrisation of the amplitude, as required by Bose-Einstein statistics. Squaring the amplitude yields the two-particle correlation:

$$R_2(p_1, p_2) = R_2(Q) = 1 + \cos(Q\Delta x), \tag{3.64}$$

with the four-momentum difference $Q = p_1 - p_2$ and the space-time difference of the sources $\Delta x = x_1 - x_2$. One obtains an enhancement of particles produced close in phase-space. Note that the dependence of R_2 on the location of the detectors dropped out. The correlation function therefore contains also information about the source of the bosons. In more realistic calculations, the particle source is modelled with various shapes and may also move in space-time [47].

In the LEP analyses [50–53], the PYBOEI model implemented in the PYTHIA Monte Carlo generator [54] is used to describe BEC effects with a Gaussian parameterisation of the correlation function:

$$R_2(Q) = 1 + \lambda e^{-Q^2 R^2}. \tag{3.65}$$

The parameter λ is the BE correlation strength, while the parameter R corresponds to the radius of the source. The correlations alter the momentum distribution of the final state particles, but the total momentum must be conserved. This is achieved by a local energy compensation with a negative BE enhancement with 1/3 of the radius R. This compensation is further constrained to vanish at $Q = 0$ by introducing an additional $1 - e^{Q^2 R^2/4}$ factor. This corresponds to the BE_{32} option of the PYTHIA generator. The LEP experiments have tuned their hadronisation models including BEC to hadronic data taken at the Z resonance, where BEC was clearly established [55] in $\pi^{\pm}\pi^{\pm}$ and $\pi^0\pi^0$ data samples.

The correlation function R_2 is determined from data by taking the ratio of the measured and background corrected two-particle density, $\rho_2(Q) = 1/N_{\text{pairs}} \frac{dN_{\text{pairs}}}{dQ}$ to the two-particle density of a reference Monte Carlo simulation without BEC, $\rho_0(Q)$:

$$R_2(Q) = \frac{\rho_2(Q)}{\rho_0(Q)} \tag{3.66}$$

Using this method, a good agreement between BEC in hadronically decaying W bosons in qq$\ell\nu$ events and hadronic Z boson decays is observed, as can be seen in Fig. 3.22. The selected Z decays are depleted in their $Z \to b\overline{b}$ content to resemble the W decay branching fractions, which mainly decay to ud and cs quark pairs.

In fully hadronic W-pair decays, correlations may also appear between final state particles from different W bosons. It would have an impact on the measurement of the mass of the W boson because it may lead to momentum transfer between the two hadronically decaying W's.

This interesting scenario of BEC between two different W bosons (inter-W BEC) is studied by comparing the two-particle density in qqqq events, $\rho_2^{WW}(Q)$, with those expected for an event sample without interference between the two decaying W bosons. This correlation function is constructed in the following way:

$$\rho_2^{WW}(Q) = 2\rho_2^{W}(Q) + 2\rho_{\text{mix}}^{WW}(Q), \tag{3.67}$$

where $\rho_2^{W}(Q)$ is the background-corrected density measured in semi-leptonic events and $\rho_{\text{mix}}^{WW}(Q)$ the density constructed by mixing uncorrelated particle pairs of independent semi-leptonic events. The mixed events are carefully arranged so that their

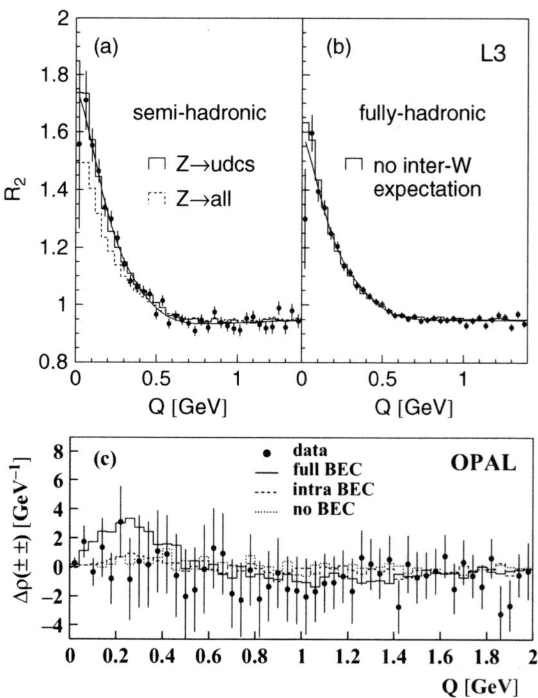

Fig. 3.22 (**a**) Two-particle correlations measured in semi-hadronic W boson decays compared to Z boson decays with and without dedicated selection of decays to light-quarks. The W and light-quark-Z distributions agree very well. (**b**) Comparison of the two-particle correlation in fully hadronic W decays compared to prediction without inter-W BEC [52]. (**c**) Difference $\delta\rho(Q)$ of the two-particle correlations measured in fully hadronic events with the correlation expected from an event sample in which inter-W BEC are absent [53]

kinematic properties agree with the qqqq event sample, including the corresponding event selection procedure.

In absence of interference between the W bosons, Eq. (3.67) holds and the difference

$$\delta\rho(Q) = \rho_2^{WW}(Q) - 2\rho_2^{W}(Q) - 2\rho_{\mathrm{mix}}^{WW}(Q) \tag{3.68}$$

is expected to vanish. Figure 3.22 shows the measured $\delta\rho(Q)$ distribution for the OPAL experiment. The agreement with zero can be further quantified by determining the integral of the $\delta\rho(Q)$ distribution from 0 to Q_{max}:

$$J = \int_0^{Q_{\mathrm{max}}} \delta\rho(Q)\, dQ. \tag{3.69}$$

Sensitivity to genuine correlations between the W bosons (inter-W correlations) is obtained in the inter-source correlation function $\delta_I(Q) = \delta\rho(Q)/\rho_{\mathrm{mix}}^{WW}(Q)$, as well as from the ratio

$$D(Q) = \frac{\rho_2^{WW}(Q)}{2\rho_2^{W}(Q) + 2\rho_{\mathrm{mix}}^{WW}(Q)}. \tag{3.70}$$

To correct possible distortions of the ratio due to the event mixing procedure or detector effects, the double ratio

$$D'(Q) = \frac{D(Q)_{\mathrm{data}}}{D(Q)_{MC,\mathrm{no-inter}}} \tag{3.71}$$

is introduced, where $D(Q)_{MC,\mathrm{no-inter}}$ is derived from a Monte Carlo sample without BEC between W bosons. $D'(Q)$ can then be parameterised as

$$D'(Q) = (1 + \delta Q)(1 + \Lambda e^{-k^2 Q^2}), \tag{3.72}$$

where Λ corresponds to the correlation strength. In the measurement of like-sign final state hadrons, the L3 experiment finds values compatible with no inter-W correlations [52]:

$$\begin{aligned} J(\pm,\pm) &= 0.03 \pm 0.33 \pm 0.15, \text{ and} \\ \Lambda &= 0.008 \pm 0.018 \pm 0.012 \end{aligned}$$

where the first uncertainty is statistical and the second systematic. The variation of track and event selection, as well as the mixing procedure contribute most to the systematic uncertainty. As a consistency check, the J integral for unlike-sign pairs yields $J(+,-) = 0.01 \pm 0.36 \pm 0.16$, consistent with zero. The expectations for a scenario with inter-W BEC are $J(\pm,\pm) = 1.38 \pm 0.10$ and $\Lambda = 0.098 \pm 0.008$,

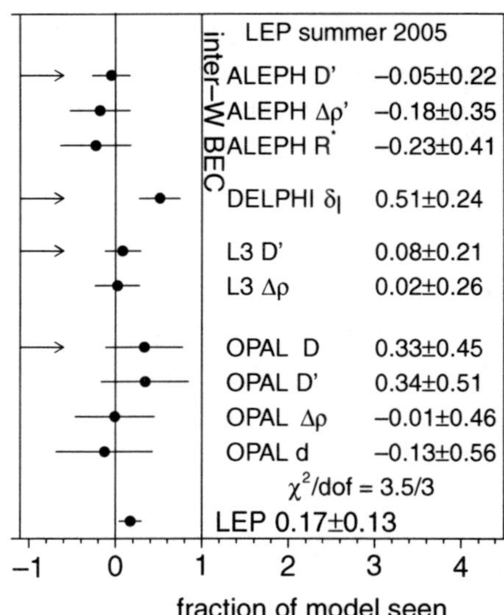

Fig. 3.23 LEP measurements of inter-W BEC and their combination [8] in terms of fraction of the BE_{32} PYBOEI model

respectively, which means the inter-W BEC model disagrees with data by more than 3.5 standard deviations.

Figure 3.23 summarises the various LEP measurement on BEC, where various variables and methods are compared in terms of the relative fraction of the inter-W BEC model that is seen in data. Combining the results of all LEP experiments, this fraction is found to be 0.17 ± 0.13 [8]. This means data are consistent with the absence of inter-W BEC and significant BEC effects according to the BE_{32} model are disfavoured.

3.6.2 Colour Reconnection

Colour reconnection (CR) is summarising QCD interference effects which lead to strong interaction between two colourless states. This phenomenon is observed, for example, in the colour-suppressed decay of B mesons, $B \to J/\psi + X$, where the c and $\bar{c}$ quarks of the $c\bar{c}$ colour singlet state of the J/ψ meson are each originating from separate colour singlet states, which have to be colour reconnected. It is therefore interesting to study if this effect is also present in hadronically decaying W's, whose decay products are necessarily colour singlets. Due to the short but non-zero lifetime of the W, the decays are separated in space. The distance is however much shorter than the typical hadronic interaction length of 1 fm, so that CR effects may alter the two W decays.

Like inter-W BEC, CR strongly affects the W mass measurement in fully hadronic W-pair events because the reconstructed mass of the W bosons would not be identical to their propagator mass. At LEP, this was the initial motivation to search for signatures of CR. The W mass is actually the observable that is most affected by CR, and without refinement of the data analysis it leads to significant systematic uncertainties on M_W in the fully hadronic decay channel.

The description of CR [56, 57] in perturbative QCD is considering pure gluon exchange between the quarks from the W decays. But the perturbative effect is suppressed by the probability $1/N_c^2$ for accidental colour reconnection. Furthermore, the momenta, k, of primary gluons generated by the reconnected system is expected to be limited by the finite width of the W boson to $k \leq k_{\max} \approx \Gamma_W M_W / E_W$. It was indeed found that the effect on the rate of charged final state hadrons, on their momentum spectrum and their distribution in phase space is only small. Not far from the W-pair threshold the CR effects on the particle-flow distribution is expected to be [58]:

$$\frac{\Delta N^{CR}}{N_{\text{no}-CR}} \leq \frac{\alpha_s(\Gamma_W)}{N_c^2} \frac{N'_q\left(k_{\max}^{CR}\right)}{N'_q(M_W/2)} \approx \mathcal{O}(10^{-2}), \tag{3.73}$$

where

$$N'_q(E) = \frac{dN_q(E)}{d\log E} \tag{3.74}$$

with the multiplicity, $N_q(E)$, inside a QCD jet of energy E. Therefore, only few low-energy particles are affected by perturbative CR.

Ad-hoc models [57–62] for the non-perturbative phase predict, however, a much larger influence on the hadronic observables, in particular on the reconstructed W mass in qqqq events. Several models are interfaced to the standard Monte Carlo generators that are used to simulate the fragmentation and hadronisation phase of the hadronic W decays.

The Sjöstrand-Khose model SK I [57] is the most popular CR model since it has a free parameter, k_I, to adjust the CR probability. The model is implemented in the Pythia program and is based on string fragmentation concepts. In the SK I model, the colour field strength, $\Omega_0(\mathbf{x}, t, \mathbf{u})$, of strings stretched out along the unit vector $\mathbf{u}$, are given an extension in phase-space with a cylindrical shape:

$$\Omega_0(\mathbf{x}, t, \mathbf{u}) = \exp\left(\frac{-(\mathbf{x}^2 - (\mathbf{ux})^2)}{2r_{\text{had}}^2}\right)\theta(t - |\mathbf{x}|)\exp\left(\frac{-(t^2 - (\mathbf{ux})^2)}{2\tau_{\text{frag}}^2}\right), \tag{3.75}$$

with the radial extension of the string r_{had} and proper time scale of the string fragmentation, $\tau_{\text{frag}} \approx 1.5$ fm$/c \approx 3r_{\text{had}}/c$. When the string is produced at space-time coordinate $(t_0, \mathbf{x}_0)$ and is propagating with $\boldsymbol{\beta}$ perpendicular to $\mathbf{u}$ the colour field is travelling according to

$$\Omega(\mathbf{x} - \mathbf{x}_0, t - t_0, \boldsymbol{\beta}, \mathbf{u}) = \Omega_0((\mathbf{x} - \mathbf{x}_0)', (t - t_0)', \mathbf{u}) \tag{3.76}$$

with the boosted coordinates $(\mathbf{x} - \mathbf{x}_0)'$ and $(t - t_0)'$.

The reconnection occurs when two colour fields of the two $W^\pm$ bosons have sufficient overlap. In general, many strings are present in each fragmenting $W^\pm$, but only the one with maximum $\Omega^\pm$ value is assumed to contribute to the reconnection probability. The calculation of the overlap integral

$$I^\pm = \int d^3x\, dt\, \Omega^+_{\max}\left(\mathbf{x}, t, \mathbf{x}_0^+, t_0^+\right) \Omega^-_{\max}\left(\mathbf{x}, t, \mathbf{x}_0^-, t_0^-\right) \tag{3.77}$$

is however time-consuming if performed numerically during the Monte Carlo generation of each event. It is therefore approximated by the sampling sum

$$I^\pm = \frac{1}{N_{\text{samp}}} \sum_{N_{\text{samp}}} \frac{\Omega^+_{\max}\Omega^-_{\max}}{\Omega_{\text{trial}}} \tag{3.78}$$

with a corresponding trial distribution

$$\Omega_{\text{trial}} = \exp\left(\frac{-\mathbf{x}^2}{f_r^2 r_{\text{had}}^2}\right) \exp\left(\frac{-t'^2}{f_t^2 \tau_{\text{frag}}^2}\right) \theta(t'), \tag{3.79}$$

with $t' = t - \max(t^-, t^+)$. The parameters $N_{\text{samp}} = 1000$, $f_r = 2.5$ and $f_t = 2.0$ are chosen to yield sufficient efficiency and numerical accuracy. When two strings overlap they are eventually reconnected according to the probability:

$$p_{\text{reco}} = 1 - e^{-k_I f_r^3 f_t I^\pm}, \tag{3.80}$$

with the free parameter k_I. At most one reconnection per event is allowed. Figure 3.24 shows the reconnection probability obtained from a reference Monte Carlo sample of WW → qqqq events at $\sqrt{s} = 189$ GeV as a function of k_I.

The models SK II and SK II' apply a string picture that corresponds to vortex lines like in type II superconductors. The topological information is concentrated in the core region around the string and reconnection occurs with unit probability when two of these regions overlap. The transverse extent of the strings can thus be neglected, which was not the case in the SK I model. There is also no free parameter.

The Ariadne program introduces CR in the dipole-cascade model [58]. The two quark pairs of the $W^\pm$ decays form two colour dipoles from which two cascades start. A randomly chosen colour index in the range 1–9 is assigned to each colour dipole in every step of the cascades. If two dipoles happen to have the same index they are allowed to reconnect. Special care is taken so that no unphysical colour flows appear. The actual reconnection is eventually taking place when the string related λ measure is reduced. This measure is defined as

$$\lambda = \sum_{1}^{n-1} \log \frac{(p_i + p_{i+1})^2}{m_0^2} \tag{3.81}$$

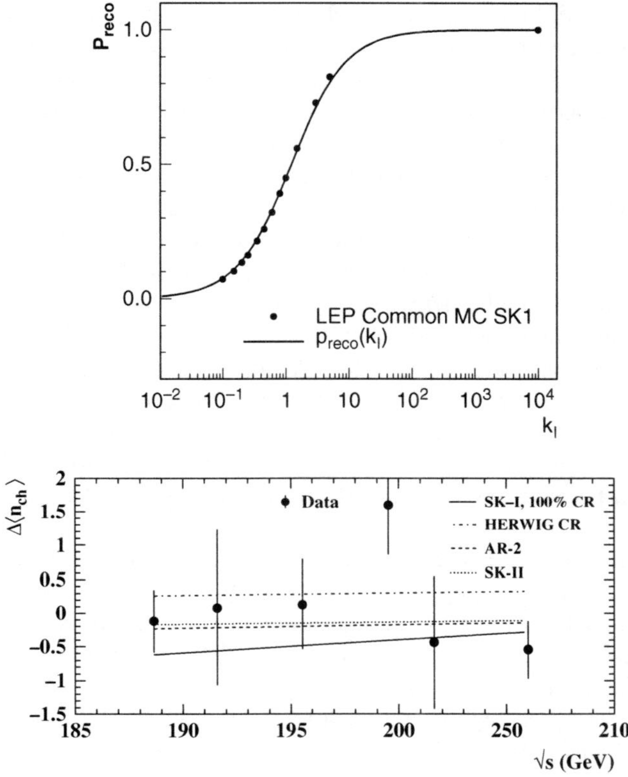

Fig. 3.24 (**a**) Colour reconnection probability as a function of the SK I model parameter, k_I, together with an approximative curve $p_{\text{reco}}(k_I) = \frac{k_I}{k_I+b}$ with $b = 1.22$. (**b**) Difference of the mean charged particle rates in fully hadronic and semi-leptonic W pair events, $\Delta\langle n_{ch}\rangle = \langle n_{ch}^{qqqq}\rangle - 2\left\langle n_{ch}^{qq\ell\nu}\right\rangle$ for different LEP centre-of-mass energies, as measured by the OPAL experiment [66]

with the four momenta p_i of the n partons along the colour string, and the mass scale $m_0 \approx 1$ GeV. This condition corresponds to the reasonable assumption that partons close in phase-space are more likely to reconnect.

In W pair events the two dipole cascades first evolve independently from large gluon energies down to $E_g \approx \Gamma_{\text{W}}$. Only in the low energy regime $E_g < \Gamma_{\text{W}}$ inter-W colour reconnection between the two cascades is turned on. The matching of the two stages of the cascade is properly taken care of. In the Ariadne AR 2 model, the inter-W reconnection probability is about 22% at a centre-of-mass energy of 189 GeV. To compare with the Ariadne model with only intra-W CR, AR 1, also this cascade is run in two stages, first down to $E_g \approx \Gamma_{\text{W}}$ and then for lower gluon energies.

Colour reconnection is as well implemented in the Herwig cluster fragmentation model [62]. In the limit of a large number of colours, every quark or anti-quark produced in the parton shower has a unique colour-connected partner with which it can be clustered. In addition, every gluon has a colour and an anti-colour index, each uniquely connected to another parton, and after showering the gluons split into quark pairs. Each quark pair is eventually forming a cluster with its colour connected partner. Like in perturbative QCD, local colour reconnection can then occur with the probability $1/N_c^2 = 1/9$, which also defines the inter-W reconnection rate.

Some of the models can also be tested at the Z peak in three-jet event topologies, where two jets originate mainly from $b\overline{b}$ pairs and the third from gluon radiation. This allowed the gluon jet to be identified with an anti-b-quark tag. In the subsequent analysis gluon jets with a rapidity gap were studied with respect to the total charged particle multiplicity [63, 64]. Similarly, the asymmetry in particle-flow between the inter-quark jet region and the quark-gluon region was measured [65]. Both type of analyses are sensitive to a rearrangement of the colour flow in the event. All LEP measurements show that the Ariadne model type 1 [58] is not consistent with data. Also the Rathsman/GAL model [61] with default parameter settings fails to describe the measurements. Other models like SK I and AR 2 could not be tested because they predict CR effects only in WW decays, not in $q\overline{q}(\gamma)$ events.

The measurement of CR in WW decay data was first investigated by analysing charged particle rates in qqqq events compared to $qq\ell\nu$ events. The OPAL experiment determined, for example, the mean number of charged particles in the two decay classes. Well reconstructed tracks in the OPAL jet chamber with a minimal transverse momentum of 0.15 GeV pointing to the interaction vertex were used in the analysis. Using all high-energy data at $\sqrt{s} = 189 - 209$ GeV the following values were found [66]:

$$\begin{aligned}
\left\langle n_{ch}^{\mathrm{qqqq}} \right\rangle &= 38.74 \pm 0.12 \pm 0.26 \\
\left\langle n_{ch}^{\mathrm{qq}\ell\nu} \right\rangle &= 19.39 \pm 0.11 \pm 0.09 \\
\Delta \langle n_{ch} \rangle = \left\langle n_{ch}^{\mathrm{qqqq}} \right\rangle - 2\left\langle n_{ch}^{\mathrm{qq}\ell\nu} \right\rangle &= -0.04 \pm 0.25 \pm 0.16 \,,
\end{aligned}$$

where the first uncertainty is statistical and the second systematic, mainly originating from uncertainties on the hadronisation model and on the comparison of $e^+e^- \to q\overline{q}$ data to simulations, which were used to refine the detector simulation. The CR models predict the values of

$$\Delta \langle n_{ch} \rangle = \begin{cases} -0.42 & (\mathrm{SK\ I},\ \mathrm{k_I} = 100) \\ -0.29 & (\mathrm{SK\ I},\ \mathrm{k_I} = 0.9) \\ -0.14 & (\mathrm{SK\ II}) \\ -0.19 & (\mathrm{AR\ 2}) \\ +0.32 & (\mathrm{HerwigCR}) \end{cases} \tag{3.82}$$

The mean value differences are all compatible with OPAL data, which is also visible in Fig. 3.24. Thus, the charge multiplicity is not sensitive enough to decide on the validity of the CR models.

A much more sensitive variable is constructed from the particle-flow [67, 68] in the qqqq event, based on the string fragmentation picture [69]. It leads to the expectation that CR changes the fragmentation in the regions between the primary quark jets. Hadronic activity between jets from different W bosons should be slightly enhanced, while the one between jets coming from the same W boson should be reduced. Furthermore, CR should mainly alter the low-momentum jet particles.

In the L3 analysis [70], a well defined sub-sample of the qqqq events is selected by requiring that the quark-jet association is optimal. The two largest inter-jet angles are required to be between 100° and 140° and not adjacent. The two other inter-jet angles must be less than 100°. In this way, the two strings between the $q\overline{q}'$ pairs evolve in opposite directions. The selection efficiency is low, in the order of 15%, but the event sample has a high purity of 85%. The rate of correct pairing of the jets to two W bosons is estimated to be 91%.

In the second step of the analysis, a plane is defined by the direction of the most energetic jet (jet 1) and the direction of the closest jet that has an angle larger than 100° with respect to jet 1. These jets are likely to originate from the same W. Since the events are in general not planar, three more planes are defined, spanned by the directions of each other jet pair 2+3, 3+4, 4+1. All particles, i, are projected on the planes and for those in the inter-jet region an angle $\phi_{i,j}$ with respect to jet j is calculated. Eventually, the analysis uses the rescaled angles

$$\phi_i^{\text{resc}} = j - 1 + \frac{\phi_{i,j}}{\psi_{j,j+1}}, \tag{3.83}$$

where $\psi_{j,j+1}$ is the angle between jet j and jet $j+1$. Figure 3.25 shows the rescaled angular distribution after background subtraction. The directions of the jets are at angles 0, 1, 2, and 3. By construction, the regions $A \in [0, 1]$ and $B \in [2, 3]$ contain preferentially particles originating from the same W boson, while the regions $C \in [1, 2]$ and $D \in [3, 4]$ contain particles which are between jets of different W bosons. To enhance the CR sensitivity, a ratio of the integrated particle-flow distributions is built:

$$R_N = \int_{0.2}^{0.8} \frac{1}{2}\left(\frac{dn_A}{d\phi} + \frac{dn_B}{d\phi}\right) \left\{\int_{0.2}^{0.8} \frac{1}{2}\left(\frac{dn_C}{d\phi} + \frac{dn_D}{d\phi}\right)\right\}^{-1} \tag{3.84}$$

The ratio of the non-integrated distributions is shown in Fig. 3.25. The ratio R_N depends on the centre-of-mass energy and for the combination of all data the values are scaled back to $\sqrt{s} = 189$ GeV. Possible reconnection effects in the remaining ZZ→ qqqq background is neglected. L3 obtains [70]:

$$R_N = 0.915 \pm 0.023\,(\text{stat.}) \pm 0.021\,(\text{syst.}). \tag{3.85}$$

The systematic uncertainty is mainly due to the experimental definition of the energy flow objects, where calorimetric cluster and track based analyses are

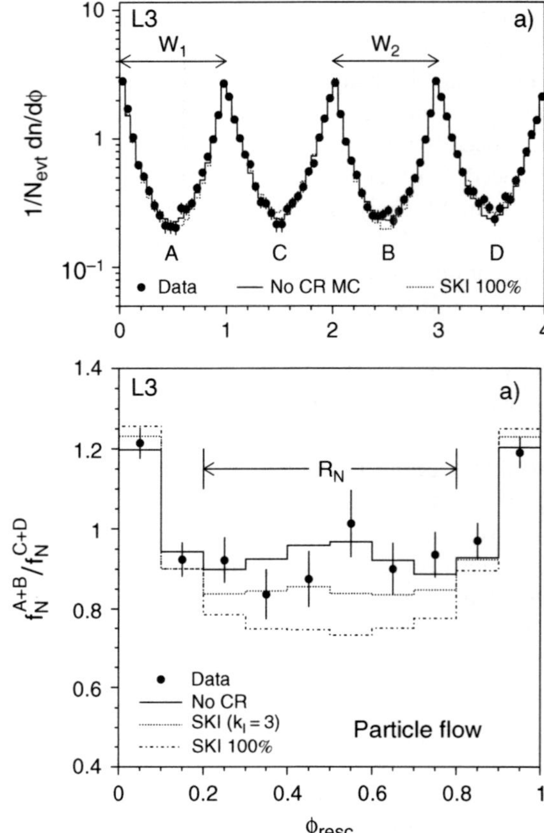

Fig. 3.25 (**a**) Distribution of the rescaled angle of the particles with respect to the next jets, as measured by the L3 experiment [70]. (**b**) Ratio of the particle-flow in the inter-jet regions A+B and C+D, compared to the standard Pythia Monte Carlo and scenarios with colour reconnection [70]

compared. The measurement agrees well with the expectation from the Pythia fragmentation without CR, $R_N(\text{Pythia, no} - \text{CR}) = 0.918 \pm 0.003$.

Similar analyses are performed by the DELPHI and OPAL experiments [66, 71]. The DELPHI analysis is based on selecting W-pair events with a particular jet configuration, like the L3 data sample, while OPAL applies an event selection used in the W mass analysis. The OPAL event selection has therefore a higher efficiency of 86% but accepts jet topologies with a more complicate colour flow. Since the measured values of R_N are not corrected for detector acceptance, resolution or efficiency the following ratio is constructed [8]:

$$r^{\text{data}} = \frac{R_N^{\text{data}}}{R_N^{\text{no-CR}}}, \tag{3.86}$$

where $R_N^{\text{no-CR}}$ is a reference value from a Monte-Carlo sample without CR. The measured r^{data} can then be compared to a model prediction, for example the $r(k_I)$ determined for the SK I model, which depends on the reconnection parameter k_I as shown in Fig. 3.26.

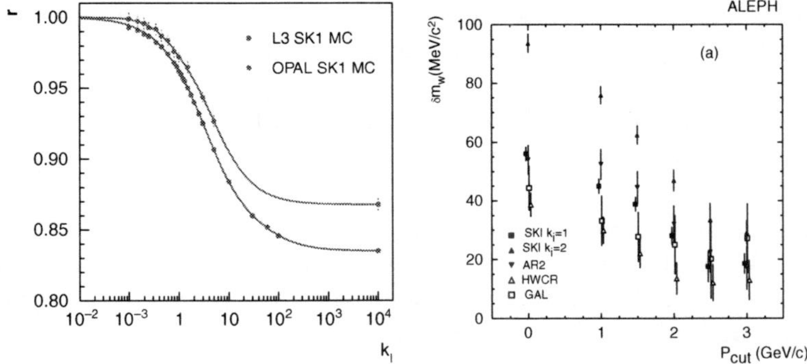

Fig. 3.26 (**a**) Evolution of the particle-flow ratio $r = R_N^{\mathrm{CR}}/R_N^{\mathrm{no-CR}}$ with the SKI parameter k_I determined by L3 and OPAL in Monte Carlo simulations. (**b**) Systematic shift of the reconstructed W mass in qqqq events due to CR effects, as a function of the momentum cut that is applied to particles in the jet reconstruction [72]

As mentioned above, the W mass measured in the LEP detectors is also a very sensitive variable to CR. In a series of studies, the effect of CR on M_{W} was found to be reduced if low-momentum tracks and low-energy cluster are removed in the jet reconstruction. Only those above a certain p_{cut} or E_{cut} are considered. Alternatively, the jets are restricted to a certain angular cone, and the new jet direction and momentum is determined in an iterative procedure. Excluding particles in the inter-jet regions outside the cones reduces the effect of CR on M_{W}. A third method applies weights, $w = p^{\kappa}$, to the momentum contribution of each particle in the jet, again suppressing low-momentum tracks, and therefore CR effects, if $\kappa > 0$, but enhancing CR effects if $\kappa < 0$.

Figure 3.26 shows the shift in M_{W} that is expected to be induced by CR to the nominal W mass value as a function of the particle momentum cut applied in the jet reconstruction. As can be clearly seen in this ALEPH study [72], the systematic shift of M_{W} is much reduced by this method. It is also remarkable that all CR models show the same trend. However, at the same time the jet resolution is worsening the more particles are removed from the jets, and so is the statistical uncertainty on M_{W} in the qqqq channel. Eventually, the best method and its corresponding cut value must be found in an optimisation procedure taking statistical and systematic uncertainties and their correlations into account. More details are given in the mass measurement section.

Important for the measurement of CR is the fact that, if CR exists, the measured $M_{\mathrm{W}}^{\mathrm{qqqq}}$ should vary when the jet reconstruction parameters are varied. The ALEPH experiment measured the variation of M_{W} as a function of p_{cut} and of the cone radius R, and a linear fit yields values compatible with no M_{W} shift:

$$\Delta M_{\mathrm{W}} = (-11 \pm 16)\,\frac{\mathrm{MeV}}{\mathrm{GeV}}\,p_{\mathrm{cut}}$$
$$\Delta M_{\mathrm{W}} = (+9 \pm 19)\,\frac{\mathrm{MeV}}{\mathrm{rad}^{-1}}\,\frac{1}{R}\,.$$

DELPHI compared the standard M_W value with alternative estimators applying a cone cut at $R = 0.5$ rad and a momentum cut at 2 GeV:

$$\Delta M_W(\text{std}, R = 0.5 \text{ rad}) = (59 \pm 35 \text{ (stat.)} \pm 14 \text{ (syst.)}) \text{ MeV}$$
$$\Delta M_W(\text{std}, p_{\text{cut}} = 2.5 \text{ GeV}) = (143 \pm 61 \text{ (stat.)} \pm 29 \text{ (syst.)}) \text{ MeV}$$

The former is well consistent with zero, the latter, however, differs from no shift by about two standard deviations. And finally, the OPAL experiment uses their most sensitive estimator, the M_W difference between the mass reconstructed applying a p_{cut} of 2.5 GeV and applying a negative momentum weight with a κ value of – 0.5:

$$\Delta M_W(p_{\text{cut}} = 2.5 \text{ GeV}, \kappa = -0.5) = (-152 \pm 68 \text{ (stat.)} \pm 61 \text{ (syst.)}) \text{ MeV}. \quad (3.87)$$

The OPAL result is compatible with zero at the 1.5 standard deviation level. All experiments performed a systematic cross-check using semi-leptonic events, in which the same jet variations are applied. The $\text{qq}\ell\nu$ analyses all gave results consistent with no effect. The observed shifts are compared to the SK I model predictions, as it is shown in Fig. 3.27 for the OPAL result.

The LEP data is eventually combined by constructing a total $\Delta\chi^2$ function which takes correlated uncertainties between the individual measurements into account. Sources of systematic uncertainties are from hadronisation and BEC effects, as well as from the modelling of the background scale and shape. The previously mentioned limit on the BEC strength is used to constrain the BEC systematics. The SK I model is taken as the main reference and k_I is the main parameter varied in the χ^2 minimisation. Figure 3.28 shows the resulting curves of the four LEP experiments together with the combined LEP measurement using information from the particle-flow analysis and the M_W-shift studies. The best value for the parameter k_I is found to be

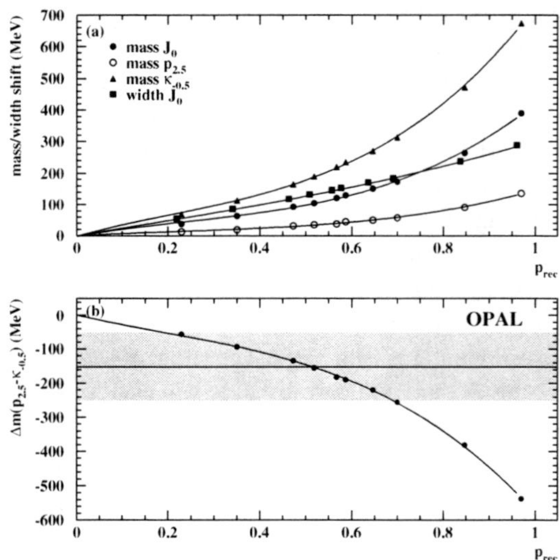

Fig. 3.27 Expected M_W shift for the SK I model as a function of reconnection probability p_{reco} compared to the OPAL ΔM_W measurement, shown as a *grey band*. The overlap region indicates the p_{reco} values preferred by data [66]

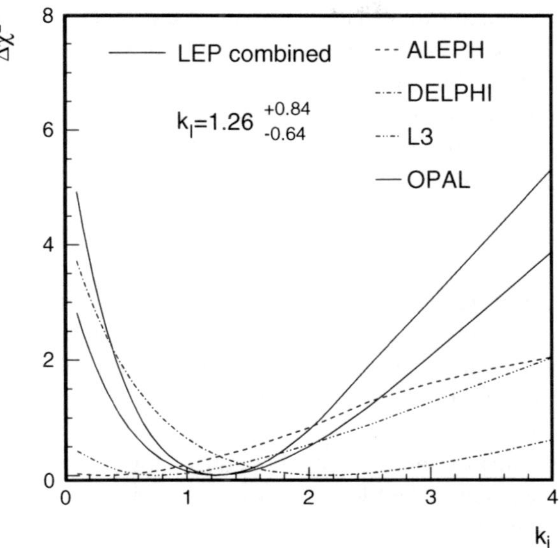

Fig. 3.28 Combined likelihood for k_I as measured in the particle-flow and M_W shift analyses of the LEP experiments, together with their combination. Systematic uncertainties and their correlations are taken into account

$$k_I = 1.26^{+0.84}_{-0.64} \tag{3.88}$$

$$k_I \in [0.62, 2.10] \quad (68\% \text{ C.L.}). \tag{3.89}$$

This corresponds to a preferred reconnection probability of 51% in the SK I model, evaluated at a centre-of-mass energy of 189 GeV. The absence of CR can not be excluded, but it is disfavoured with 2.8 standard deviations. On the other hand, an extreme CR scenario with 100% reconnection fraction is ruled out with a significance of 6.9 standard deviations. The LEP data therefore supports the existence of CR as modelled in SK I. Other CR scenarios like AR 2 and Herwig predict smaller reconnection probabilities of 22 and 11%, respectively. In dedicated studies of these two CR models by the LEP experiments [72, 71, 70, 66] both show only small deviations from the no-CR scenario, which is not favoured by the combined LEP result. When re-interpreting the SK I measurement, the AR-2 model agrees with data at the 2.0 standard deviation level, while for Herwig the consistency is at the 2.4 standard deviation level.

The measurement of CR in hadronic W-pair events is an interesting physics result by itself because it confirms the existence of colour rearrangement as observed in colour-suppressed meson decays. The reconnection probability is even beyond the naive expectation when simply counting the number of colour combinations, $1/N_c^2$. The determination of CR effects directly from data is a very important ingredient for the W-mass measurement at LEP. As discussed below, CR is one of the main sources of systematics on M_W and the result is used to estimate the corresponding uncertainties.

The CR measurement at LEP may also be helpful for physics at $p\bar{p}$ and pp colliders, Tevatron and LHC. In the simulation of the underlying event CR effects play a significant role [73]. The CR Monte-Carlo parameters are usually tuned to measured event shape variables of the underlying event and thus constrained [74]. The CR models used in these estimations are implemented in Pythia [54] and typically assume a 33% direct reconnection probability with nearest neighbours in momentum space. New multiple interaction (MI) models like Colour Annealing [75] are defining a probability for a given string to not participate in the reconnection process as:

$$p_{\text{keep}} = (1 - \eta)^{n_{\text{MI}}} \tag{3.90}$$

where η is a free parameter and n_{MI} the number of interactions that occurred in each event, which increases the reconnection probability for any given string in events with many interactions. In addition, the model tends to perform colour connection to the partons closest in momentum space, and therefore minimising the total string length. This model would in principle be valid also in WW decays, but it was not studied by the LEP collaborations. However, it may well be directly compared to the result in the SK I framework. This may give an additional handle to control CR systematics at Tevatron and the LHC. This is of importance for the measurement of the top quark mass, where the CR uncertainty currently amounts to 0.41 GeV, being among the three largest single contributions to the total systematic uncertainty on m_t [76].

3.7 Measurement of the W Boson Mass

The mass of the W boson, M_W, is a central parameter in the Standard Model, and a precise determination of M_W is one of the main tasks of the LEP experiments. There are two methods used to measure this quantity at LEP: by determining the W-pair production rate at the threshold and by direct reconstruction of the W decay spectrum. Since the LEP physics goals were not only the determination of W parameters but also the searches for new particles, only a fraction of the total data, about $4 \times 20\ \text{pb}^{-1}$, were recorded at threshold energies. The largest data samples of about $4 \times 680\ \text{pb}^{-1}$, are available at energies between 183 and 209 GeV.

At the W pair threshold, the cross-section of W pair production is dominated by t-channel neutrino exchange and it is proportional to the velocity of the W bosons:

$$\sigma_{WW} \propto \beta = \frac{p_W}{E_W} = \sqrt{1 - \frac{4M_W^2}{s}} \tag{3.91}$$

The first determination of M_W at LEP was therefore derived from the cross-section measurement at the optimal $\sqrt{s}$ value of 161 GeV, and at 172 GeV. They yield a W-mass value of [8]:

$$M_W = 80.40 \pm 0.20\,(\text{stat.}) \pm 0.07\,(\text{syst.}) \pm 0.03\,(E_{\text{beam}})\ \text{GeV} \tag{3.92}$$

which has a rather large uncertainty, dominated by the statistical uncertainty on σ_{WW}. In principle, M_W could have been measured more precisely if more data would have been collected in the threshold energy range. With the same amount of data, a precision very similar to the one from fully reconstructed events can be reached, with different systematic uncertainties. This method may therefore be used again at a future linear e^+e^- colliders [5]. As mentioned earlier, an improved accuracy of the theoretical predictions for σ_{WW} at the threshold is being worked on [77].

In the direct reconstruction method, the masses of each decaying W is determined from the measured leptons and jets. All LEP experiments analysed the $qq\ell\nu$ and qqqq final states, OPAL also determined M_W in $\ell\nu\ell\nu$ events.

In the fully leptonic final state the W masses can not be completely reconstructed because of the two neutrinos of the W decay. However, neglecting the lepton masses and the finite W width the dependence of lepton energy, E_ℓ, on M_W is given by

$$E_\ell = \frac{\sqrt{s}}{4} + \cos\theta_\ell^* \sqrt{\frac{s}{16} - \frac{M_W^2}{4}}, \tag{3.93}$$

where $\cos\theta_\ell^*$ is the angle between the lepton momentum in the W rest frame and the direction of the W in the laboratory frame. This angle is not measurable, and the sensitivity of the lepton energy to M_W is mainly from the endpoints of the spectrum, where $\cos\theta_\ell^* = \pm 1$. When assuming that the neutrinos are in the same plane as the leptons the event kinematics can be solved with respect to a pseudo-mass, $M_\pm$, up to a two-fold ambiguity. The solutions are [78]:

$$\begin{aligned} M_\pm^2 = \frac{2}{|\mathbf{p}_{\ell'} + \mathbf{p}_\ell|^2} \Big\{ & (P\mathbf{p}_{\ell'} - Q\mathbf{p}_\ell)(\mathbf{p}_{\ell'} + \mathbf{p}_\ell) \\ & \pm \sqrt{|\mathbf{p}_{\ell'} \times \mathbf{p}_\ell|^2 [|\mathbf{p}_{\ell'} + \mathbf{p}_\ell|^2 (E_{\text{beam}} - E_\ell)^2 - (P+Q)^2]} \Big\}, \end{aligned} \tag{3.94}$$

with $P = E_{\text{beam}} E_\ell - E_\ell^2 + \frac{1}{2} m_\ell^2$, $Q = E_{\text{beam}} E_{\ell'} - \mathbf{p}_{\ell'} \mathbf{p}_\ell + \frac{1}{2} m_{\ell'}^2$, the beam energy E_{beam} and the lepton masses and momenta, $\mathbf{p}_{\ell/\ell'}$ and $m_{\ell/\ell'}$. OPAL performed a likelihood analysis fitting parameterised spectra to the data distributions of the two quantities. The fit results in

$$M_W(\ell\nu\ell\nu) = 80.41 \pm 0.41 \pm 0.13 \text{ GeV}, \tag{3.95}$$

where the first uncertainty is statistical and the second systematic. Here, the beam energy uncertainty, QED radiative corrections, as well as background modelling are taken into account. The main systematic uncertainties are due to the lepton momentum scale, which is determined from $Z \to \ell\ell$ events, both at the Z peak and at higher energies. Although the precision is not extraordinary, the analysis of this decay channel is important because systematic effects from hadronisation and FSI are completely absent.

Precise knowledge of the lepton momentum scale and resolution is also important when extracting M_W in the qqeν and qq$\mu\nu$ channels. The hadronically decaying W can be fully reconstructed from the two quark jets. Using momentum conservation, also the neutrino four-momentum can be calculated and thus the four-momentum of the leptonically decaying W. Final state photon radiation (FSR) is predominantly emitted along the charged lepton or quark jet and included in the quark jets and reconstructed leptons. Photons radiated from the initial state electrons (ISR) are detected in about 5% of the events as isolated clusters in the calorimeters. When identified, the photon clusters are excluded in the formation of the jets.

The resolution for the W masses is improved by applying a kinematic fit to the event. The measured lepton energies and angles, as well as the jet energies and directions are varied within their resolution until energy-momentum conservation is fulfilled. The variation of the jet momenta (or jet masses) in the kinematic fit is done by keeping the jet velocity β (or the boost γ) of the jets constant because many systematic effects cancel in the corresponding ratios with the jet energy. Since the momentum conservation was already exploited to calculate the neutrino momentum, this results in a fit with one constraint (1C). Furthermore, a second constraint is applied requiring the two W masses in the event to be equal within the W width. Figure 3.29 shows an example of the L3 data analysis, where the mass resolution is reduced by the 2C fit by about a factor of two in the qqeν and qq$\mu\nu$ channels. Information form both 1C and 2C masses are usually used in the subsequent mass analyses.

In the qq$\tau\nu$ channel, the kinematic constraints are spoilt by the additional neutrino from the τ decay. Only the hadronically decaying W boson contains mass information. The mass resolution can however be improved by applying a rescaling of the sum of the jet energies to the beam energy, where a factor two can be gained in resolution. Overlap of the leptonic τ decays with the qqeν and qq$\mu\nu$ channels is avoided by applying strict separation cuts, for example on the $M_{\ell\nu}$ mass.

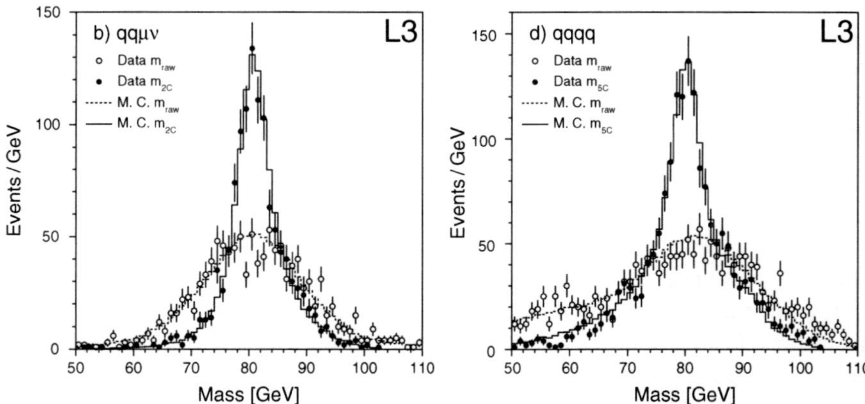

Fig. 3.29 Improvement in mass resolution by applying a kinematic fit in semi-leptonic and fully hadronic W-pairs [79]

In the fully hadronic channel the complete final state can be reconstructed, including isolated ISR photons in the forward detectors. QCD gluon radiation is taken into account by splitting the event samples in four-jet and five-jet events, using, for example, the Durham [27] jet resolution parameter y_{45} as a measure to separate the two topologies. Energy-momentum conservation in the kinematic fit corresponds to four constraints (4C) and the equality of the W masses adds another constraint (5C). The improvements in mass resolution are in the order of a factor two to three, for the 4C and 5C fits, as illustrated in Fig. 3.29. Like in the $qq\ell\nu$ case, information from both fit classes are typically used to measure M_W.

The four and five jets in the qqqq$(+g)$ events can be paired in three, respectively ten, ways to form two W decays. They are distinguished by ordering the jets in energy, so that a probability for each combination can be calculated. In the 4-jet case all three combinations have at least a 5–10% probability to be correct and they are all considered in the mass analysis, assuming that the combinatorial background is well described by Monte Carlo. Figure 3.30 gives an impression of the size of the combinatorial background with respect to the correctly reconstructed signal and the real background from non-WW decays. In 5-jet events, some combinations have a negligible probability to be correct, e.g. the case in which the second most energetic jet is combined with the least energetic one. The ALEPH experiment selects only one combination in their analyses using a pairing probability that is based on the CC03 matrix element calculated for the reconstructed jets [72]. The other experiments use a W mass estimator that combines all pairings that have a high probability to be correct. They are weighted accordingly in the combined mass likelihood. The weights are based on the polar angle of the reconstructed W boson, the sum of jet charges of each jet combination and the transverse momentum of the gluon jet in 5-jet events (DELPHI [80]), the probability of the kinematic fit (L3 [80]), or a neural network variable (OPAL [81]) trained with the reconstructed mass differences as input.

The extraction of M_W and Γ_W from the reconstructed mass spectra is performed with various methods. ALEPH and L3 apply a Monte Carlo template method in which the measured spectra are compared as 2- or 3-dimensional distributions to Monte Carlo samples with different underlying M_W and Γ_W values. The test statistics for the data to Monte Carlo comparison is either a unbinned likelihood (L3) or binned histograms (ALEPH), where the binning is optimised to obtain a bias-free measurement. The unbinned likelihood is, for example, constructed as [79]

$$L(M_W, \Gamma_W) = \prod_{i=0}^{N_{\text{data}}} \frac{1}{\sigma_s(M_W, \Gamma_W) + \sigma_b} \left\{ \frac{d\sigma_s\left(M_W, \Gamma_W, m_1^i, m_2^i\right)}{dm_1\, dm_2} + \frac{d\sigma_s\left(m_1^i, m_2^i\right)}{dm_1\, dm_2} \right\}, \tag{3.96}$$

where σ_s and σ_b are the signal and background cross-sections, and m_1 and m_2 the mass estimators, like the 1C and 2C, or the 4C and 5C mass pairs. ALEPH uses in addition the uncertainty on the 2C and 5C masses as a third variable. The likelihoods are evaluated for each decay channel and each centre-of-mass energy separately. The differential cross-sections is calculated from Monte Carlo by the sum of template events in an interval I in the (m_1, m_2) space close to the measured $\left(m_1^i, m_2^i\right)$ point of event i, divided by the size of the interval ΔI and normalised to the Monte Carlo luminosity L_{MC}:

$$\frac{d^2\sigma}{dm_1\, dm_2} \approx \frac{1}{L_{MC}} \sum_I \frac{1}{\Delta I}. \tag{3.97}$$

The Monte Carlo templates are of large statistics, usually 10^6 events, to reduce statistical fluctuations in the calculation of $d^2\sigma/(dm_1\, dm_2)$. If the Monte Carlo describes all detector effects and physics phenomena correctly, the W parameters can then be extracted without any bias.

The generation of templates with all M_W and Γ_W values that are needed to perform the likelihood maximisation is however impossible, in the sense that it would take a lot of computing time. The virtue of the template method is therefore in the reweighting of the Monte Carlo samples. A weight is attributed to each simulated event j according to the ratio of the matrix element squared:

$$w_j = \frac{\left|M\left(p_k^j, M_W, \Gamma_W\right)\right|^2}{\left|M\left(p_k^j, M_W^{MC}, \Gamma_W^{MC}\right)\right|^2} \tag{3.98}$$

with the M_W and Γ_W values that are to be determined and the nominal M_W^{MC} and Γ_W^{MC} values of the original Monte Carlo sample. The matrix element also depends on the four-momenta p_k^j of the generated final state fermions of event j, and possibly on the four-momenta of ISR or FSR photons. The matrix elements are calculated using four-fermion Monte Carlo programs, like EXCALIBUR [40] or

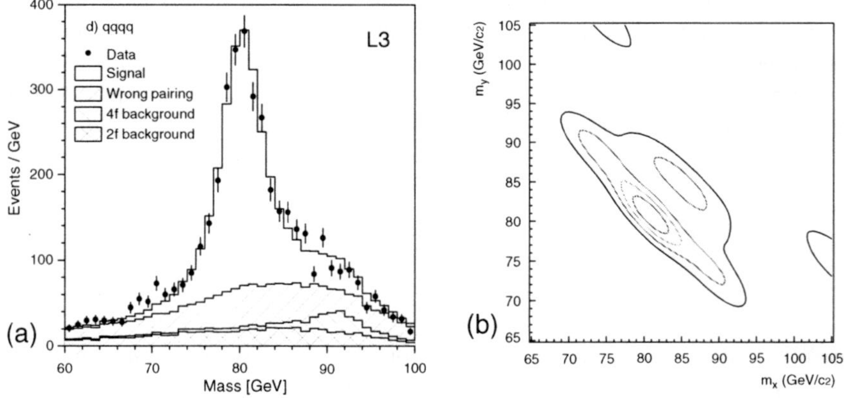

Fig. 3.30 Reconstructed W mass spectrum in fully hadronic events with combinatorial, 2-fermion and 4-fermion background (*left*). Combined ideogram of a 4-jet W-pair event, as analysed by DELPHI (*right*) [80]

KORALW/KandY [29]. The extraction of the correct W mass and width value from a given data set is verified in tests with high statistics Monte Carlo samples and also using many samples, each with the expected size of the data set.

The OPAL experiment uses a convolution fit to determine M_W and Γ_W from data [81]. A normalised physics function P describes the double-differential cross-section in the two reconstructed W masses m_1 and m_2 :

$$P(m_1, m_2, M_W, \Gamma_W) = a_0 \left\{ B(m_1, M_W, \Gamma_W) \otimes B(m_2, M_W, \Gamma_W) \otimes S\left(m_1, m_2, \sqrt{s'}\right) \right\} \otimes I(\sqrt{s}, \sqrt{s'}) \tag{3.99}$$

with the normalisation factor a_0, the Breit-Wigner distributions

$$B(m, M_W, \Gamma_W) = \frac{m^2}{\left(m^2 - M_W^2\right)^2 + (m^2 \Gamma_W / M_W)^2} \tag{3.100}$$

a phase space term

$$S(m_1, m_2, \sqrt{s'}) = \sqrt{(s' - (m_1 + m_2)^2)(s' - (m_1 - m_2)^2)} \tag{3.101}$$

and the radiator function

$$I(\sqrt{s'}, \sqrt{s}) = \beta x^{\beta - 1} \frac{\sigma(\sqrt{s'}, M_W)}{\sigma(\sqrt{s}, M_W)}, \tag{3.102}$$

which describes the reduction in centre-of-mass energy from $\sqrt{s}$ to $\sqrt{s'}$ due to ISR with energy fraction $x = E_\gamma / \sqrt{s}$ and $\beta = (2\alpha_{\mathrm{QED}}/\pi) \log((\sqrt{s}/m_e)^2 - 1)$.

The physics function P is then folded with the resolution function R to obtain the likelihood L_i^s of each event i to be compatible with the WW signal:

$$L_i^s(M_W, \Gamma_W) = R_i(m_1, m_2) \otimes P(m_1, m_2, M_W, \Gamma_W) \tag{3.103}$$

The total likelihood is taking also the parameterised background likelihood into account:

$$L_i = p_i^s L_i^s(M_W, \Gamma_W) + p_i^b L_i^b, \tag{3.104}$$

where the two likelihood terms are weighted with event-by-event probabilities to be signal or background, p_i^s and p_i^b. In case of qqqq events, information of each jet pairing is entering the likelihood as a separate estimator, weighted by the corresponding neural network output.

The DELPHI analysis [80] applies the convolution technique in the qq$\ell\nu$ channels, similar to OPAL. In the qqqq channel, however, DELPHI exploits a priori the most of the information in each event. The idea is to convolute the predicted probability density $P(m_1, m_2, M_W, \Gamma_W)$ with the probability density of the complete 4C fit. The latter is however difficult to compute in a time-efficient way for the whole (m_1, m_2) space. An approximation is therefore applied. The 4C χ^2 function is evaluated at the minimum of the (m_1, m_2) space together with the covariance matrix V between m_1 and m_2. With this information a new χ^2 is constructed as:

$$\chi^2(m_1, m_2) = \chi^2_{4C,\text{min}} + (\mathbf{m} - \mathbf{m}^{\text{min}})V^{-1}(\mathbf{m} - \mathbf{m}^{\text{min}})^T \tag{3.105}$$

with $\mathbf{m} = (m_1, m_2)$ and $\mathbf{m}_{\text{min}} = \left(m_1^{\text{min}}, m_2^{\text{min}}\right)$. From this, the 4C probability density is easily calculated

$$P_{4C}(m_1, m_2) = e^{-\frac{1}{2}\chi^2(m_1, m_2)} \tag{3.106}$$

The probability density is determined for all possible jet pairings, for three different jet clustering algorithms (Durham [27], Cambridge [82], Diclus [83]), and assuming an ISR photon escaping along the beam direction in the kinematic fit. A weighted sum of the 18 so-called ideograms is calculated in the 4-jet case, and 60 weighted ideograms are used in the 5-jet case. Figure 3.30 gives an example of these ideograms for a fully hadronic 4-jet event. Although technically complicated, the method obtained very good linearity when comparing fitted with underlying M_{W} values in large statistics Monte Carlo samples. A global mass bias, in the order of 200 MeV with an uncertainty of less than 10 MeV, is observed and being corrected for.

When only the W mass is a free parameter in the fit, the W width is assumed to follow the Standard Model relation:

$$\Gamma_{\text{W}} = \frac{3G_{\text{F}}M_{\text{W}}^3}{2\sqrt{2}\pi}\left(1 + 2\frac{\alpha_s}{3\pi}\right). \tag{3.107}$$

With the very refined measurement techniques, each experiment reaches a statistical precision on M_{W} between 54 and 70 MeV in the combined $\text{qq}\ell\nu$ channels and between 59 and 70 MeV in the qqqq channel. The precision on the latter is however reduced by applying a globally optimised jet reconstruction to account for FSI effects. The differences between the results of the LEP collaborations are mainly due to intrinsic experimental differences, like acceptance, resolution, and detection efficiencies.

The systematic error sources can be subdivided into correlated and uncorrelated systematics. Correlations can exist between analysis channels, between measurements at different centre-of-mass energies, and between experiments.

Each experiment studied the lepton energy scale and linearity, the angular measurement, and their resolutions in great detail. Usually, two-fermion events at the Z peak and at higher energies, measured in the same data taking periods as the W pair events, are used to determine remaining differences between Monte Carlo simulation and data. These differences were corrected for and their uncertainties translated into systematic uncertainties in the lepton and jet measurements. Effects that were rather negligible in any previous measurement at LEP became important, for example the detailed distribution of calorimetric clusters close to leptons which influences the hadronic mass in $qq\ell\nu$ events. Also the alignment across sub-detectors and their relative orientation to the beam were checked to avoid an angular bias. In the combined LEP measurement, the lepton systematics enter as correlated only between different data sets. Jet systematics are in addition correlated between the $qq\ell\nu$ and qqqq channels, but uncorrelated between experiments. In total, the detector uncertainties contribute with 10 MeV in the $qq\ell\nu$ channel and 8 MeV in the qqqq channel.

Backgrounds are mainly from Z pair and two-fermion production, whose cross-sections are also measured directly. The corresponding uncertainties are used to scale the background distributions globally by common factors over the whole mass spectrum. Also the slope of the background contributions in the measured spectra is changed. All experiments verified the background distributions in independent samples and apply missing corrections to the Monte Carlo predictions. In the LEP combination, this uncertainty is combined with the uncertainty due to limited Monte Carlo statistics and contributes with 3 MeV to the error in the $qq\ell\nu$ measurement, where backgrounds small, and with 11 MeV to the qqqq error.

An uncertainty common to all measurements is the determination of the LEP beam energy. Since the beam energy constraint is applied in the kinematic fit, the beam energy error translates directly into an error on M_W. The W width is much less affected, also by the beam energy spread. The beam energy was determined by the LEP energy working group at each interaction point with time intervals of 10 min. Using the time information the correct beam energy is thus calculated for each individual W pair event. Data were grouped in different centre-of-mass energy bins, for which a global energy calibration was applied. This calibration is based on the flux-loop, beam spectrometer and tune shift measurements, as described in Chap. 2. When combining all data, the complete correlation matrix between these energy bins is used, resulting in an 8 MeV uncertainty on the LEP M_W value.

Photon radiation evidently influences the reconstructed W mass spectra. The combined Monte Carlo programs KORALW and YFSWW3 (KandY) includes ISR effects in the YFS exponentiation scheme to $O(\alpha^3)$, full $O(\alpha)$ electroweak corrections, including interference between ISR, FSR and W radiation (WSR), as well as screened Coulomb corrections. The latter describe Coulomb interactions between the W bosons, which are potentially large but screened due to the limited lifetime of the W bosons. Higher order, leading-log FSR corrections are included using PHOTOS for leptons and Pythia for quarks. Alternatively, the RacoonWW Monte Carlo is used, which is also based on the complete $O(\alpha)$ matrix element completed with an ISR radiator function. ISR effects are generally estimated by comparing the $O(\alpha^3)$ with the $O(\alpha^2)$ calculation, yielding small shifts of about 1 MeV on M_W [28, 30, 29]. The effect of Coulomb screening is studied by taking half of the difference between Monte Carlo samples with screened Coulomb effect and without any Coulomb effect, which amounts to about 7 MeV. To study the uncertainty on the non-leading $O(\alpha)$ electroweak corrections, a good estimate is derived from the direct comparison of the RacoonWW and the KandY generators. Some care has to be taken in this comparison since collinear photons are not explicitly generated in RacoonWW. The observed differences are in order of 10 MeV for $qq\ell\nu$ and 5 MeV for qqqq. Alternatively, the uncertainty on the $O(\alpha)$ corrections are derived from removing non infra-red photons radiated from W bosons and applying the $O(\alpha)$ corrections not only to the CC03 part of the event weight, but also on the difference between CC03 and the full 4f contribution. Both checks together give a 2 MeV uncertainty on M_W. Since some error estimates overlap and the experiments apply different strategies, the total LEP uncertainty due to radiative corrections is 8 MeV in the semi-leptonic channel and 5 MeV in the fully hadronic channel, assuming full correlation between all data sets [8].

Quark fragmentation and hadronisation is another common systematic error source that needs to be taken into account. The LEP experiments compare the Pythia, Herwig and Ariadne models, which are carefully tuned to Z decay events. The hadronic Z decay samples are depleted in b-quark jets, because these are practically absent in W decays at LEP. Additional attention was put on the rate of heavy hadrons produced in jets, like kaons and baryons. This is because the standard jet clustering algorithms assume hadrons with either zero mass or they attribute the pion mass to each cluster. If the rates of the more heavy hadrons is not exactly reproduced by Monte Carlo, this leads to a bias in the reconstructed W mass. In L3, for example, the baryon content of the simulation was compared to measurements [84] and good agreement was observed, as can be seen in Fig. 3.31. The uncertainty on the baryon rate translates into an systematic uncertainty on M_W and Γ_W. Similar comparisons and adjustments of the baryon rate are applied by the other collaborations [72, 80, 81], generally reducing the hadronisation uncertainty. Eventually, in the LEP combination, these uncertainties are taken as fully correlated and contribute with 13 and 19 MeV in the $qq\ell\nu$ and qqqq channel, respectively.

A systematic effect only present in the fully hadronic channel is from final state interactions, Bose-Einstein correlations (BEC) and colour reconnection (CR). The

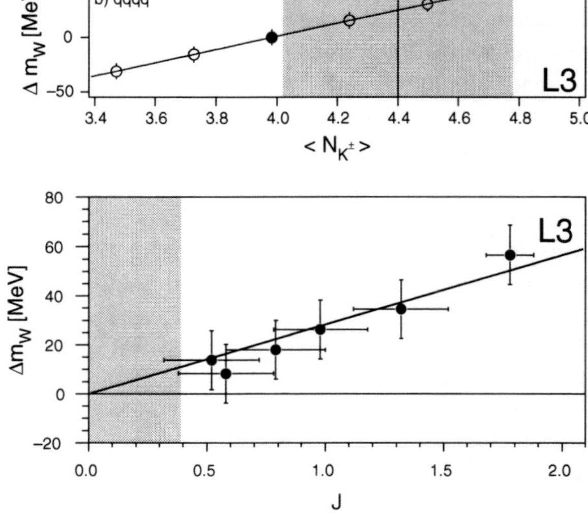

Fig. 3.31 Variation of the reconstructed W mass as function of the kaon content in qqqq events (*top*). The *grey area* indicates the measurement of the kaon rate, the *open circles* the Monte Carlo prediction, and the *full circle* the actual value used in the Monte Carlo simulation. The *bottom plot* shows the correlation of the W mass in qqqq events with the Bose-Einstein parameter J which is found to be in good agreement with a linear function [79]

measurements of these effects are described in detail in the previous sections. LEP measurements are consistent with the absence of inter-W BEC, so that the baseline W mass and width analyses assume only intra-W BEC. The systematic uncertainty on M_W is assessed by translating the BEC intra-W limit into a limit on the possible W mass shift. The BEC observables are actually found to have a linear correlation with M_W, as can be seen for the integral parameter J in Fig. 3.31. This is not trivial since J depends both on the modelled pion source radius and the correlation strength. The one standard deviation limit on the intra-W BEC strength of 30% corresponds to a 7 MeV uncertainty on M_W in the hadronic channel.

Colour reconnection turned out to be a much harder problem than BEC. The effects on M_W are potentially large in the SK I model, even using the limit $k_I < 2.10$ from the direct measurement of CR. The corresponding shift of the hadronic W mass using the default jet clustering is in the order of 90 MeV at a centre-of-mass energy of 189 GeV [72]. However, CR affects mainly the inter-jet regions and low momentum particles, so that the CR systematics is much reduced when jets are limited to the high energy particles in the jet core. As shown in Fig. 3.26 the CR mass shift is only in the order of 30 MeV if for example a momentum cut at 3 GeV is applied. Each experiment is optimising the jet reconstruction with respect to the total uncertainty in the qqqq channel. ALEPH is applying a particle momentum cut at 3 GeV, and DELPHI uses an iterative jet cone procedure with a cone radius of $R = 0.5$ rad. The optimal cluster energy cut found by L3 is at 2 GeV, while OPAL removes particles with momenta below 2.5 GeV. The differences in the optimal analyses can be explained by the limits on CR that are obtained by the different experiments. L3, for example, finds a rather low limit on k_I at 1.1, while the ALEPH limit is around 2.0. For the same $k_I < 2.0$, also the L3 analysis would select 3 GeV as the optimal working point. In the LEP combination, this leads to a reduced weight for the mass measurements which are not close to the LEP combined limit of $k_I = 2.10$. Overall, the CR uncertainty in the hadronic channel is 35 MeV, assuming CR according to SK I with the given LEP limit.[1]

The final systematic uncertainties in the LEP combined W mass and width measurements are summarised in Table 3.1. The masses in the semi-leptonic and hadronic channel are found to be [8]:

$$M_W(\mathrm{qq}\ell\nu) = 80.372 \pm 0.030\ (\mathrm{stat.}) \pm 0.020\ (\mathrm{syst.})\ \ \mathrm{GeV}$$
$$M_W(\mathrm{qqqq}) = 80.387 \pm 0.040\ (\mathrm{stat.}) \pm 0.044\ (\mathrm{syst.})\ \ \mathrm{GeV}$$

with a correlation coefficient of 0.20. The consistency of the two measurements is tested by taking the difference of the two, where CR and BEC errors are set to zero, which yields:

$$\Delta M_W(\mathrm{qqqq} - \mathrm{qq}\ell\nu) = -12 \pm 45\ \ \mathrm{MeV}, \tag{3.108}$$

[1] The systematic uncertainty on M_W due to CR is actually derived for a k_I limit of 2.13, but no numerical difference is expected with respect to the current systematics of 35 MeV.

Table 3.1 Decomposition of systematic and statistical uncertainties on the LEP combined W mass measurement [8]

	Systematic uncertainty on M_W in MeV		
Source	$qq\ell\nu$	qqqq	Combined
ISR/FSR/$O(\alpha)$	8	5	7
Hadronisation	13	19	14
Detector systematics	10	8	10
LEP beam energy	9	9	9
Colour reconnection	–	35	8
Bose-Einstein correlations	–	7	2
Other	3	11	4
Total systematic	21	44	22
Statistical	30	40	25
Total	36	59	33
Statistical in absence of systematics	30	27	20

a value well compatible with zero. This indicates that CR effects are efficiently reduced in the final LEP analyses and that they are not larger than expected from the various models. Eventually, combining all LEP data from threshold and direct reconstruction data, the W mass is found to be

$$M_W^{LEP} = 80.376 \pm 0.025\,(\text{stat.}) \pm 0.022\,(\text{syst.})\ \text{GeV}. \tag{3.109}$$

This value agrees well with the Tevatron measurement of the W mass [85] of

$$M_W^{p\bar{p}} = 80.420 \pm 0.031\ \text{GeV}, \tag{3.110}$$

and also with the W mass from the analysis of all electroweak data, excluding the direct measurement [86]

$$M_W^{EW} = 80.364 \pm 0.020\ \text{GeV}. \tag{3.111}$$

The LEP W-mass measurements are therefore still very competitive with the Tevatron results for this important Standard Model parameter.

The LEP experiments analysed the W decay spectra to also determine the W width. The same combination procedure was applied as in the W-mass measurement, including systematic uncertainties and their correlations. The W width data are not optimised with respect to the FSI effects, which are dominating the systematics together with the hadronisation effects. Taking the relatively large statistical uncertainty into account, an optimisation is however not needed. Combining all LEP data the combined fit yields:

$$\Gamma_W = 2.196 \pm 0.063\,(\text{stat.}) \pm 0.055\,(\text{syst.})\ \text{GeV}. \tag{3.112}$$

This is again in good agreement with the combined CDF and DØ measurement of [85]

$$\Gamma_{\rm W} = 2.050 \pm 0.058 \ \text{GeV}. \tag{3.113}$$

Both are consistent with the Standard Model value of $\Gamma_{\rm W} = 2.091 \pm 0.002$ GeV [86].

Summarising the results of this chapter, electroweak boson production is in very good agreement with Standard Model predictions. The measurements of vector boson couplings are confirming the non-abelian $SU(2) \times U(1)$ gauge structure of the theory. Furthermore, the LEP era provides precise measurements of the vector boson couplings to fermions and of the Z and W boson masses. A discussion about the consistency with other precision measurements within the Standard Model and possible hints on new physics follows in the next chapter.

References

1. R. Gastmans and T. T. Wu, *The Ubiquitous Photon: Helicity Methods for QED and QCD*, Oxford University Press (1990).
2. The ALEPH Collaboration, the DELPHI Collaboration, the L3 Collaboration, the OPAL Collaboration, the SLD Collaboration, the LEP Electroweak Working Group, the SLD electroweak, heavy flavour groups, Phys. Rept. **427** (2006) 257; hep-ex/0509008v3.
3. The Particle Data Group, C. Amsler, et al., Phys. Lett. **B 667** (2008) 1.
4. L3 Collaboration, P. Achard et al., Phys. Lett. **B 587** (2004) 16.
5. A. Djouadi, J. Lykken, K. Mönig, Y. Okada, M. Oreglia, S. Yamashita, et al., *ILC Reference Design Report Volume 2 - Physics at the ILC*, arXiv:0709.1893.
6. B. Odom, D. Hanneke, B. D'Urso and G. Gabrielse, Phys. Rev. Lett. **97** (2006) 030801; G. Gabrielse, D. Hanneke, T. Kinoshita, M. Nio and B. Odom, Phys. Rev. Lett. **97** (2006) 030802, Erratum, Phys. Rev. Lett. **99** (2007) 039902.
7. The L3 Collaboration, P. Achard, et al., Phys. Lett. **B 616** (2005) 145.
8. The LEP Collaborations ALEPH, DELPHI, L3, OPAL, and the LEP Electroweak Working Group, *A Combination of Preliminary Electroweak Measurements and Constraints on the Standard Model*, CERN-PH-EP/2006-042; hep-ex/0612034.
9. The OPAL Collaboration, G. Abbiendi, et al., Eur. Phys. J. **C 26** (2003) 331.
10. The DELPHI Collaboration, J. Abdallah, et al., Eur. Phys. J. **C 30** (2003) 447.
11. The L3 Collaboration, P. Achard, et al., Phys. Lett. **B 72** (2003) 133.
12. S. Jadach, G. Passarino and R. Pittau, (eds.) *Reports of the working groups on precision calculations for LEP2 Physics*, CERN Preprint CERN-2000-009.
13. S. Jadach, W. Placzek and B.F.L. Ward, Phys. Rev. **D 56** (1997) 6939.
14. E. Accomando, A. Ballestrero and E. Maina, Comput. Phys. Commun. **150** (2003) 166; hep-ph/0204052.
15. J. Fujimoto, et al., Comp. Phys. Comm. **100** (1997) 128; hep-ph/9605312.
16. The L3 Collaboration, P. Achard, et al., Phys. Lett. **B 597** (2004) 119.
17. The OPAL Collaboration, G. Abbiendi, et al., Phys.Lett. **B 604** (2004) 31.
18. S. Roth, *Precision Electroweak Physics at Electron-Positron Colliders*, Springer Tracts in Modern Physics, Vol. 7002 (2006).
19. LEP Energy Working Group, Assmann, R., et al., Eur. Phys. J. **C 6** (1999) 187; LEP Energy Working Group, Eur. Phys. J. **C 39** (2005) 253.
20. G. Altarelli,. T. Sjostrand and F. Zwirner, (eds.) *Workshop on Physics at LEP2*, Yellow report CERN 96-01, 1996.

21. K. Hagiwara, R. D. Peccei and D. Zeppenfeld, Nucl. Phys. **B 282** (1987) 253.
22. M. Bilenky, J. L. Kneur, F. M. Renard and D. Schildknecht, Nucl. Phys. **B 409** (1993) 22.
23. The ALEPH Collaboration, S. Schael, et al., Eur. Phys. J. **C 38** (2004) 147.
24. The DELPHI Collaboration, J. Abdallah, et al., Eur. Phys. J. **C34** (2004) 127.
25. The L3 Collaboration, P. Achard, et al., Phys. Lett. **B 600** (2004) 22.
26. The OPAL Collaboration, G. Abbiendi, et al., Eur. Phys. J. **C 52** (2007) 767.
27. S. Catani et al., Phys. Lett. **B 269** (1991) 432; S. Bethke et al., Nucl. Phys. **B 370** (1992) 310.
28. F. Cosutti, Eur. Phys. J. **C44** (2005) 383.
29. KandY runs concurrently KORALW version 1.51 and YFSWW3 version 1.16. S. Jadach et al., Comp. Phys. Comm. **140** (2001) 475.
30. Racoon WW, A. Denner, S. Dittmaier, M. Roth and D. Wackeroth, Phys. Lett. **B 475** (2000) 127; A. Denner, S. Dittmaier, M. Roth and D. Wackeroth, *W-pair production at future e+e- colliders: precise predictions from RACOONWW*, hep-ph/9912447.
31. The ALEPH Collaboration, S. Schael, et al., Phys. Lett. **B 605** (2005) 49.
32. The DELPHI Collaboration, J. Abdallah, et al., Eur. Phys. J. **C 45** (2006) 273.
33. The L3 Collaboration, P. Achard, et al., Phys. Lett. **B 547** (2002) 151.
34. G. Passarino, Nucl. Phys. **B 578** (2000) 3; G. Passarino, Nucl. Phys. **B 574** (2000) 451.
35. see for example: W. Kilian, *Dynamical Electroweak Symmetry Breaking*, hep-ph/0303015v2.
36. The ALEPH Collaboration, S. Schael, et al., Phys. Lett. **B 614** (2005) 7.
37. The DELPHI Collaboration, J. Abdallah, et al., Eur. Phys. J. **C 54**(2008) 345.
38. The L3 Collaboration, P. Achard, et al., Phys. Lett. **B 586** (2004) 151.
39. The OPAL Collaboration, G. Abbiendi, et al., Eur. Phys. J. **C 33** (2004) 463.
40. F. A. Berends, R. Kleiss and R. Pittau, Nucl. Phys. **B 424** (1994) 308; Nucl. Phys. **B 426** (1994) 344; Nucl. Phys. (Proc. Suppl.) **B 37** (1994) 163; R. Kleiss and R. Pittau, Comp. Phys. Comm. **83** (1994) 141; R. Pittau, Phys. Lett. **B 335** (1994) 490.
41. M. Diehl and O. Nachtmann, Z. Phys. **C 62** (1994) 397.
42. http://www-cdf.fnal.gov/physics/ewk/, public notes are regularly appearing; The DØ Collaboration, V. Abazov, et al., Phys. Rev. Lett. **100** (2008) 131801, arXiv.org:0712.0599; The DØ Collaboration, V. Abazov, et al., Phys. Lett. **B 653** (2007) 378, arXiv.org:0705.1550; The DØ Collaboration, V. Abazov, et al., Phys. Rev. **D 74** (2006) 057101, hep-ex/0608011; The DØ Collaboration, V. Abazov, et al., *Measurement of the WW production cross section with dilepton final states in p-pbar collisions at sqrt(s)=1.96 TeV and limits on anomalous trilinear gauge couplings*, arXiv:0904.0673v1; The DØ Collaboration, V. Abazov, et al., Phys. Rev. **D 76** (2007) 111104, arXiv.org:0709.2917.
43. The ALEPH Collaboration, A. Heister, et al., Phys. Lett. **B 602** (2004) 31; The DELPHI Collaboration, J. Abdallah, et al., Eur. Phys. J. **C 31** (2003) 139; The L3 Collaboration, P. Achard, et al., Phys. Lett. **B 540** (2002) 43; The L3 Collaboration, P. Achard, et al., Phys. Lett. **B 527** (2002) 29; The L3 Collaboration, P. Achard, et al., Phys. Lett. **B 490** (2000) 187; The L3 Collaboration, P. Achard, et al., Phys. Lett. **B 478** (2000) 39; The OPAL Collaboration, G. Abbiendi, et al., Phys. Rev. **D 70** (2004) 032005; The OPAL Collaboration, G. Abbiendi, et al., Phys. Lett.**B 471** (1999) 293.
44. G. Bélanger and F. Boudjema, Nucl. Phys. **B 288** (1992) 201; J. W. Stirling and A. Werthenbach, Eur. Phys. J. **C 14** (2000) 103.
45. R. Hanbury Brown and R. Q. Twiss, *A Test of a New Type of Stellar Interferometer on Sirius*, Nature **178** (1956) 1046.
46. G. Goldhaber, et al., Phys. Rev. Lett. **3** (1959) 181.
47. D. H. Boal, C. -K. Gelbke and B. K. Jennings, Rev. Mod. Phys. **62** (1990) 553.
48. G. Baym, Act. Phys. Pol. **B 29** (1998) 1839.
49. J. Vandalen, *Bose-Einstein correlations in e+e- events*, Ph.D. thesis, University of Nijmegen (2002).
50. The ALEPH Collaboration, S. Schael, et al., Phys. Lett. **B 606** (2005) 265.
51. The DELPHI Collaboration, J. Abdallah, et al., Eur. Phys. J. **C 44** (2005) 161.

52. The L3 Collaboration, P. Achard, et al., Phys. Lett. **B 547** (2002) 139.
53. The OPAL Collaboration, G. Abbiendi, et al., Eur. Phys. J. **C 36** (2004) 297.
54. T. Sjöstrand, P. Edén, C. Friberg, L. Lönnblad, G. Miu, S. Mrenna and E. Norrbin, Comp. Phys. Comm. **135** (2001) 238; For colour reconnection and Bose-Einstein correlation studies PYTHIA version 6.121 is used; T. Sjöstrand, *Recent Progress in PYTHIA*, Preprint, LU-TP-99-42, hep-ph/0001032.
55. The L3 Collaboration, P. Achard, et al., Phys. Lett. **B 524** (2002) 55.
56. G. Gustafson, U. Pettersson and P. Zerwas, Phys. Lett. **B 209** (1988) 90.
57. T. Sjöstrand and V. Khoze, Z. Phys. **C 62** (1994) 281; V. Khoze and T. Sjöstrand, Eur. Phys. J. **C 6** (1999) 271.
58. L. Lönnblad, Z. Phys. **C 70** (1996) 107.
59. G. Gustafson and J. Häkkinen, Z. Phys. **C 64** (1994) 659.
60. J. Ellis and K. Geiger, Phys. Rev. **D 54** (1996) 1967, Phys. Lett. **B 404** (1997) 230.
61. J. Rathsman, Phys. Lett. **B 452** (1999) 364.
62. G. Corcella, I. G. Knowles, G. Marchesini, S. Moretti, K. Odagiri, P. Richardson, M. H. Seymour and B. R. Webber, JHEP **0101** (2001) 010; hep-ph/0210213.
63. The ALEPH Collaboration, S. Schael, et al., Eur. Phys. J. **C 48** (2006) 685.
64. The OPAL Collaboration, G. Abbiendi, et al., Eur. Phys. J. **C 35** (2004) 293.
65. The L3 Collaboration, P. Achard, et al., Phys. Lett. **B 581** (2004) 19.
66. The OPAL Collaboration, G. Abbiendi, et al., Eur. Phys. J. **C 45** (2006) 291.
67. D. Duchesneau, *New Method Based on Energy and Particle Flow in* $e^+e^- \rightarrow WW \rightarrow$ *hadron Events for Colour Reconnection Studies*, LAPP-EXP-2000-02.
68. A. Ballestrero, et al., J. Phys. **G 24** (1998) 365.
69. The JADE Collaboration, W. Bartel and A. Petersen, *Results From Jade*, Talks given at the 15th Rencontre de Moriond, Les Arcs, France, Mar 9–21, 1980; The JADE Collaboration, W. Bartel, et al., Phys. Lett. **B 101** (1981) 129; The JADE Collaboration, W. Bartel, et al., Z. Phys. **C 21** (1983) 37; The JADE Collaboration, W. Bartel, et al., Phys. Lett. **B 134** (1984) 275; The JADE Collaboration, W. Bartel, et al., Phys. Lett. **B 157** (1985) 340; The TPC/Two Gamma Collaboration, H. Aihara, et al., Z. Phys. **C 28** (1985) 31; The TPC/Two Gamma Collaboration, H. Aihara, et al., Phys. Rev. Lett. **57** (1986) 945; The TASSO Collaboration, M. Althoff, et al., Z. Phys. **C 29** (1985) 29.
70. The L3 Collaboration, P. Achard, et al., Phys. Lett. **B 561** (2003) 202.
71. The DELPHI Collaboration, J. Abdallah, et al., Eur. Phys. J. **C 51** (2007) 249.
72. The ALEPH Collaboration, S. Schael, et al., Eur. Phys. J. **C 47** (2006) 309.
73. C. Buttar, et al., *Les Houches Physics at TeV Colliders 2005, Standard Model and Higgs working group: Summary report*, arXiv:hep-ph/0604120v1.
74. R. Field, et al., *Pythia Tune A, Herwig, and Jimmy in Run 2 at CDF*, hep-ph/0510198.
75. D. Wicke, P. Z. Skands, *Non-perturbative QCD Effects and the Top Mass at the Tevatron*, arXiv:0807.3248v1.
76. Tevatron Electroweak Working Group, for the CDF Collaboration, the DØ Collaboration, *Combination of CDF and D0 Results on the Mass of the Top Quark*, arXiv:0903.2503v1
77. M. Beneke, et al., *Four-fermion production near the W pair production threshold*, CERN-PH-TH-07-107, arXiv:0707.0773v1 [hep-ph].
78. The OPAL Collaboration, G. Abbiendi, et al., Eur. Phys. J. **C 26** (2003) 321.
79. The L3 Collaboration, P. Achard, et al., Eur. Phys. J. **C 45** (2006) 569, hep-ex/0511049.
80. The DELPHI Collaboration, J. Abdallah, et al., Eur. Phys. J. **C 55** (2008) 1, arXiv:0803.2534http://arxiv.org/abs/0803.2534.
81. The OPAL Collaboration, G. Abbiendi, et al., Eur. Phys. J. **C 45** (2006) 307, hep-ex/0508060.
82. Yu. L. Dokshitzer, G. Leder, S. Moretti and B. Webber, JHEP **08** (1997) 001; S. Bentvelsen, I. Meyer, *The Cambridge Jet algorithm: features and applications*, hep-ph/9803322v1.
83. L. Lönnblad, Z. Phys. **C 58** (1993) 471.
84. The DELPHI Collaboration, P. Abreu, et al., Eur. Phys. J. **C 18** (2000) 203; Erratum ibid. **C 25** (2002) 493.

85. Tevatron Electroweak Working Group, *Combination of CDF and D0 results on the W boson mass and width*, arXiv:0808.0147v1; Tevatron Electroweak Working Group, *Updated Combination of CDF and D0 results for the Mass of the W Boson*, arXiv:0908.1374v1.
86. The ALEPH, DELPHI, L3, OPAL, SLC Collaborations, the LEP Electroweak Working Group, the Tevatron Electroweak Working Group, and the SLD electroweak and heavy flavour groups, *Precision Electroweak Measurements and Constraints on the Standard Model*, CERN-PH-EP/2008-020; arXiv:0811.4682. Updates can be found at http://lepewwg.web.cern.ch/LEPEWWG/.

Chapter 4
Electroweak Measurements and Model Analysis of Electroweak Data

The parameters of the Standard Model are accessed in many precision measurements at the LEP, SLD, and Tevatron colliders and in low-Q^2 experiments. The low-Q^2 range is meant to be relative to the square of the weak boson masses, M_Z^2 and M_W^2.

As mentioned in Chap. 1, of main interest are the electroweak coupling structure and the particle masses, especially those of the heavy particles, m_t, M_W, M_Z, and M_H. In the following sections the different inputs to the combined analysis of electroweak data are discussed.

4.1 W Boson Mass Measurements at LEP and Tevatron

The mass of the W boson is not only measured at LEP but also in $p\bar{p}$ collisions at the Tevatron. W bosons are produced in the parton-level process $q\bar{q}' \to W$ and decays to $e\nu_e$ and $\mu\nu_\mu$ are selected. The observables sensitive to M_W [1] are the transverse lepton momentum p_T^ℓ, the missing transverse momentum $p_T^{\text{miss}} = p_T^\nu$, and the transverse mass m_T, which is an approximation of the mass of the decaying W. It is calculated according to

$$m_T = \sqrt{2 p_T^\ell p_T^{\text{miss}} (1 - \cos \Delta\phi)} , \tag{4.1}$$

with the azimuthal angular difference, $\Delta\phi$, between the missing momentum and p_T^ℓ. Examples of the measured m_T and p_T^ℓ distributions in the $W \to e\nu$ channel are shown in Fig. 4.1. The W mass is extracted from a binned log-likelihood fit to the m_T, p_T^ℓ and p_T^{miss} spectra, in which the data in each bin is compared to predictions using different underlying M_W values, so-called templates. In case of the CDF measurement, the templated predictions are determined in a fast simulation procedure. The likelihoods are scanned in M_W steps of 1 MeV. The minima are determined for each data set and decay channel separately and the results are eventually combined, taking correlations into account.

The m_T spectrum yields statistically and systematically the most precise M_W value, while the p_T^ℓ and p_T^{miss} measurements have an about 20% larger error. The

A. Straessner, *Electroweak Physics at LEP and LHC*, STMP 235, 111–135,
DOI 10.1007/978-3-642-05169-2_4,

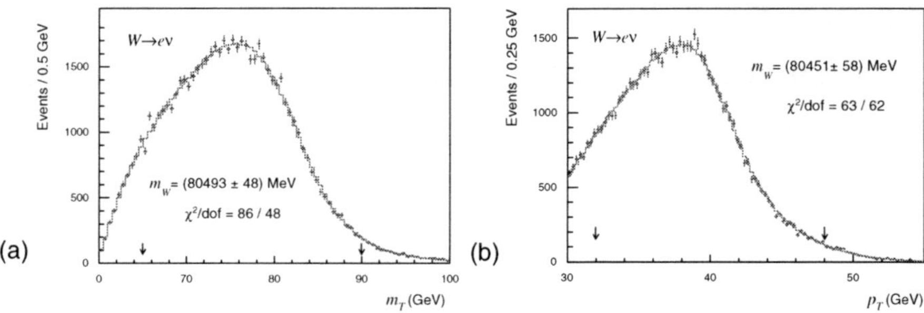

Fig. 4.1 Transverse mass (**a**) and transverse lepton momentum (**b**) distribution in $W \to e\nu$ events as measured by CDF [2]. The fit result using the two spectra in the range indicated by the *arrows* are also shown in the graphic

main sources of systematic uncertainties is from the tracker momentum scale in the muon and electron channel and the calorimeter energy scale for electrons. The former is calibrated in di-muon decays of J/ψ, Υ and Z bosons, while the latter relies on the precise calibration of the E/p ratio of electrons. It is also verified with $Z \to \mathrm{e}^+\mathrm{e}^-$ events. In the latest CDF measurement, the lepton scale contributes with 23 MeV to the total systematic uncertainty of 34 MeV. Furthermore, uncertainties from variation of PDFs are determined with the standard CTEQ6 [3] error sets and comparison with the alternative MRST [4] parameterisation. They contribute with 13 MeV to the systematics. In the p_T^ℓ and p_T^{miss} measurements the recoil scale and resolution play a larger role than for m_T, so that the corresponding uncertainty on the combined M_W value is 8–10 MeV each. The influence of the transverse momentum of the W, p_T^W, on the measured p_T^ℓ is well controlled by fits to Drell-Yann production data, exploiting the similarity to Z boson production. It actually affects the M_W value only by 4 MeV, even less than the uncertainties from lepton resolution, efficiency, backgrounds and simulation uncertainties, which each contribute between 2 and 6 MeV. The CDF measurement analysing 200 pb^{-1} of Run-II data yields

$$M_\mathrm{W} = 80.413 \pm 0.034\ (\text{stat.}) \pm 0.034\ (\text{syst.})\ \text{GeV}. \tag{4.2}$$

The currently best DØ measurement is based on 1 fb^{-1} of Run-II data and the analysis of only the $W \to \mathrm{e}\nu$ channel yields [5]:

$$M_\mathrm{W} = 80.401 \pm 0.021\ (\text{stat.}) \pm 0.038\ (\text{syst.})\ \text{GeV}\ . \tag{4.3}$$

with a total uncertainty smaller than in the CDF determination. The main source of systematics in this measurement is from the electron momentum scale and amounts

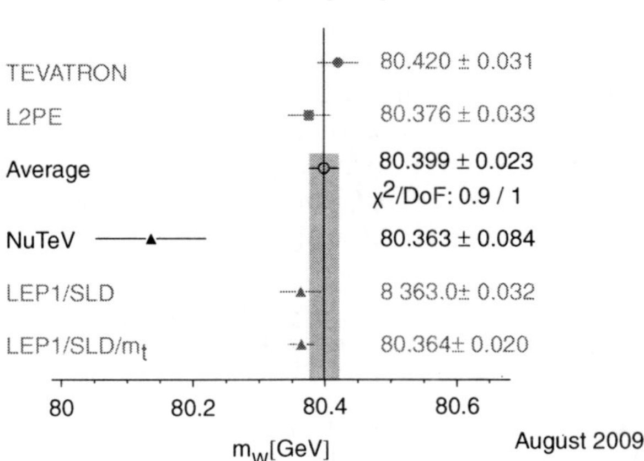

Fig. 4.2 Tevatron and LEP measurement of the W boson mass and their combination [13]

to 34 MeV. The CDF and DØ results are combined taking correlations into account, which yields [1]:

$$M_{\mathrm{W}} = 80.420 \pm 0.031 \;\; \mathrm{GeV} \tag{4.4}$$

The uncorrelated uncertainties sum up to about 27 MeV, while the main correlated uncertainties are about 12 MeV from assumptions about the W and Z boson production, respectively the parton density functions applied in the analyses, and about 9 MeV from the description of radiative corrections.

The Tevatron result on M_{W} is compared to the LEP measurement in Fig. 4.2. The currently best M_{W} value combines all collider results and yields:

$$M_{\mathrm{W}} = 80.399 \pm 0.023 \;\; \mathrm{GeV} \,. \tag{4.5}$$

It agrees well with the W-mass derived in the analysis of other electroweak data of

$$M_{\mathrm{W}} = 80.364 \pm 0.020 \;\; \mathrm{GeV} \,, \tag{4.6}$$

which mainly exploits the well-known relation between the precisely measured muon decay constant G_{F} and the electroweak gauge boson masses of Eq. (1.45). Radiative corrections and uncertainties on the other model parameters, including a variation of M_{H} between the current lower limit of 114.4 GeV and 1 TeV, are taken into account.

4.2 Top Mass Measurement at the Tevatron

The mass of the heaviest quark dominates radiative correction terms to the W propagator over other quark contributions, and a precise determination of m_{t} is therefore necessary for detailed model comparisons with electroweak data. The top quark mass is above the W+b threshold so that it practically always decays to this final state since W+s and W+d decays are CKM suppressed. The width of the top quark in the Standard Model is given at NLO by [6]:

$$\Gamma_t = \frac{G_{\mathrm{F}} m_{\mathrm{t}}^3}{8\pi\sqrt{2}} \left(1 - \frac{M_{\mathrm{W}}^2}{m_{\mathrm{t}}^2}\right)^2 \left(1 + 2\frac{M_{\mathrm{W}}^2}{m_{\mathrm{t}}^2}\right) \left\{1 - \frac{2\alpha_s}{3\pi}\left(\frac{2\pi^2}{3} - \frac{5}{2}\right)\right\} \quad (4.7)$$

which numerically is equal to 1.35 GeV, for the currently best M_{W} and m_{t} values and $\alpha_s = 0.118$. The corresponding lifetime is very short, $\tau_t = 5\times 10^{-25}$s, so that the top decays before hadronisation starts or $t\bar{t}$ bound states can be formed. Experimentally, an upper limit of 1.8×10^{-13}s is derived at 95% C.L. by reconstructing decay vertices of top quarks and their corresponding decay lengths [7].

At the Tevatron, the dominant production of top quarks is in $t\bar{t}$ pairs with the following final states:

- 46.2% fully hadronic: $t\bar{t} \to W^+bW^-\bar{b} \to q\bar{q}'bq''\bar{q}'''\bar{b}$
- 43.5% lepton+jets: $t\bar{t} \to W^+bW^-\bar{b} \to q\bar{q}'b\ell^-\bar{\nu}_\ell\bar{b} + \ell'^+\nu_{\ell'}bq''\bar{q}'''\bar{b}$
- 10.3% di-lepton: $t\bar{t} \to W^+bW^-\bar{b} \to \ell^+\nu_\ell b\ell'^-\bar{\nu}_{\ell'}\bar{b}$

The branching fractions of the three decay types are identical to the corresponding W pair decay fractions (see Sect. 3.3).

The most precise m_{t} value is obtained in the analysis of the lepton+jets channel [8, 9]. Top quark pairs are selected using the final state signatures, one lepton with high p_T^ℓ, high p_T^{miss}, two light-quarks jets and two jets with a B hadron tag which is based on the fact the B hadrons travel several millimetres before they decay. The event kinematics of the two top quarks in the event are reconstructed from the final state leptons and jets, and where all jet-to-parton permutations are taken into account exploiting also the b-tag information. One of the potentially largest systematic uncertainty is due to the jet energy scale (JES), because it enters directly the top mass estimator. The JES for light jets can however be determined in top decay events themselves by fixing the mass of the hadronically decaying W bosons to the externally measured M_{W} value. The JES for b-jets is then determined relative to the JES of light jets.

In the matrix-element weighting method, which is used in many top mass analyses, a likelihood is constructed from the event probability

$$P_e(x; m_{\mathrm{t}}, \mathrm{JES}) = f_{\mathrm{top}} P_s(x; m_{\mathrm{t}}, \mathrm{JES}) + (1 - f_{\mathrm{top}}) P_b(x; \mathrm{JES}) \ , \quad (4.8)$$

where x summarises the kinematic variables of the event, f_{top} denotes the signal fraction, and P_s and P_b the signal and background probabilities. They both depend on the event kinematics and the JES. The signal probability is sensitive to m_t and can be written as [9]:

$$P_s(x; m_t, \text{JES}) = \frac{1}{\sigma_{s,\text{obs}}(x; m_t, \text{JES})} \times \sum_i w_i \int_{q_1,q_2,y} dq_1 dq_2 f(q_1) f(q_2) \frac{(2\pi)^4 |M(q\bar{q} \to t\bar{t} \to y)|^2}{2q_1 q_2 s} W(x, y; \text{JES}) d\Phi_6 \,. \quad (4.9)$$

The matrix element squared, $|M(q\bar{q} \to t\bar{t} \to y)|^2$, is weighted by the PDFs, $f(q_i)$, and integrated over the parton momentum fractions q_i. A second integral is performed over the parton configurations y that correspond to the measured event kinematics x, where $W(x, y; \text{JES})$ is the transfer function that relates y to x and which depends also on the JES. The integral is normalised by the observed signal cross-section $\sigma_{s,\text{obs}}(x; m_t, \text{JES})$. All parton-jet permutations are weighted with w_i and combined. In this way, a maximum of information is extracted from each event and m_t and the JES can be fitted simultaneously. The fit result for DØ in the leptons+jets channel is shown in Fig. 4.3.

Residual p_T or η dependent uncertainties on the JES which can not be accessed by a global rescale factor are furthermore taken into account. Also the b-JES is further tested by varying b decay fractions, b fragmentation functions and detector response to b-jets. In the lepton+jets channel the total JES uncertainty is about 1.0 and 1.3 GeV for CDF and DØ, respectively. Other systematic error sources, like PDF uncertainties, ISR and FSR modelling, background uncertainties and modelling of the underlying event and pile-up are relatively smaller, so that the total systematics is at the 1.4–1.6 GeV level, compared to a 0.8–0.9 GeV statistical error that both Tevatron experiments obtain in about 3.2 (CDF) and 3.6 fb^{-1} (DØ) [11].

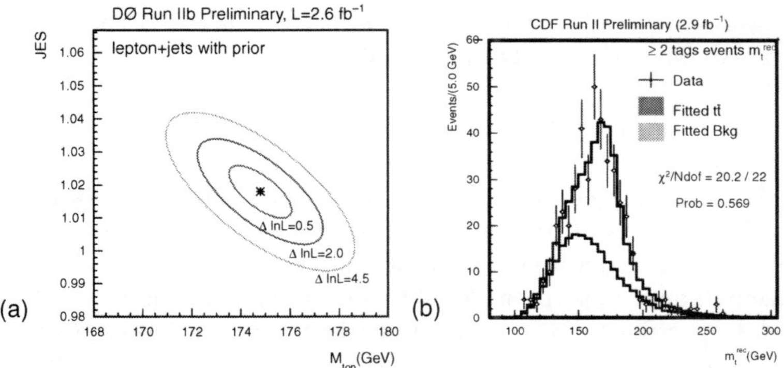

Fig. 4.3 (**a**) Correlation of extracted top mass and jet energy scale (JES) in lepton+jets event analysis of DØ [9]. (**b**) Top mass spectrum in fully hadronic top decays with two b-tagged jets, as determined by CDF [10]

The combined m_t and JES fit is also applied in the fully hadronic channel, while the di-lepton channel can obviously not rely on this technique. Here, the JES uncertainty is determined by varying jet energy scales within its uncertainty, in addition to the previously described JES systematics. The total JES error is therefore slightly larger, 2.0–2.6 GeV, than in the channels with hadronic top decays. An example of a mass spectrum measured in the hadronic channel is shown in Fig. 4.3.

The top quark mass can however not only be measured by fully reconstructing its decay into W+b. Since at the Tevatron the top quarks are nearly produced at rest, the b-quark system is boosted by a relativistic factor of

$$\gamma_b = \frac{m_t^2 + m_b^2 - M_W^2}{2 m_t m_b} \approx 0.4 \frac{m_t}{m_b}\,, \tag{4.10}$$

which depends on m_t. Rather than determining the lifetime of the B hadron its two-dimensional decay length, L_{2D}, is measured. This is the distance between the primary vertex of the event and the reconstructed secondary vertex from the B decay in the plane transverse to the beam. The mean $\langle L_{2D} \rangle$ has a nearly linear relation with m_t. Similarly, the mean p_T^ℓ of the lepton is highly correlated with the top mass because also the W boson receives a boost proportional to m_t. CDF obtains the following results in 1.9 fb^{-1} of data [12]:

$$\begin{aligned} \langle L_{2D} \rangle &= 0.596 \pm 0.017 \text{ cm } \Rightarrow m_t = 176.7^{+10.0}_{-8.9} \text{ (stat.)} \pm 3.4 \text{ (syst.) GeV} \\ \left\langle p_T^\ell \right\rangle &= 55.2 \pm 1.3 \text{ GeV } \Rightarrow m_t = 173.5^{+8.9}_{-9.1} \text{ (stat.)} \pm 4.2 \text{ (syst.) GeV} \end{aligned}$$

These measurements are by construction not directly affected by JES uncertainties. They are only indirectly influenced by a change of the event selection, mainly because of out-of-cone corrections to the JES. The total effect is however much smaller than other systematics. Dominating are the scale uncertainties on the $\langle L_{2D} \rangle$ and $\left\langle p_T^\ell \right\rangle$ and QCD ISR and FSR effects.

An overview over all top quark measurements by CDF and DØ and their combination are shown in Fig. 4.4. The central value [11] of

$$m_t = 173.1 \pm 0.6 \text{ (stat.)} \pm 1.1 \text{ (syst.) GeV} \tag{4.11}$$

is mainly influenced by the lepton+jet channel but all analyses yield very consistent results. The main systematic effects are from JES (±0.7 GeV), and Monte-Carlo modelling of the $t\bar{t}$ signal used for calibrating the fit method (±0.5 GeV). The uncertainty from the understanding of Colour Reconnection in $t\bar{t}$ events amounts to ±0.4 GeV, estimated by comparing two differently tuned Monte-Carlo parameter sets with CR. Similarly to the W-mass measurements at LEP, the CR systematics may become a dominating uncertainty since it is fully correlated among experiments and analysis channels, and because the JES uncertainty will decrease

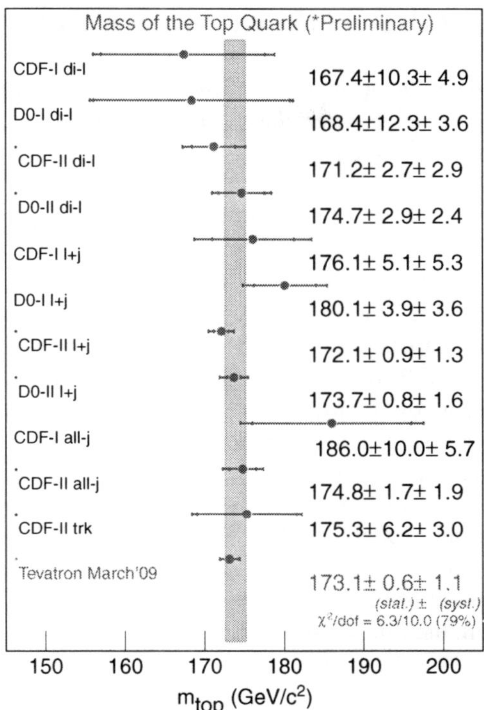

Fig. 4.4 Tevatron measurements of the top quark mass and their combination [11]

with increasing data statistics. The uncertainties on signal and background description are each contributing with ±0.3 GeV, which takes ISR, FSR and hadronisation effects into account. The remaining systematics from lepton momentum scale, multi-hadron interactions and finite Monte Carlo statistics are all smaller than 0.2 GeV.

A comparison with the indirect top quark mass of $m_{\mathrm{t}} = 178.9^{+11.7}_{-8.6}$ GeV [13] determined from other electroweak data is only a rough test of the Standard Model relations between the electroweak parameters, which emphasises the important role of a precise direct measurement of the top quark mass.

4.3 Low-Q^2 Measurements

In the general analysis of electroweak data, precision experiments at energies below the Z resonance are taken into account as well. Atomic parity violation, Møller scattering and Neutrino-nucleon scattering are discussed in the following. The hadronic cross-section measurement plays a separate, but important role, because it is input to the determination of the running of the electromagnetic coupling α_{QED} from $Q^2 = 0$ to $Q^2 = M_Z^2$, where many LEP and SLD measurements were performed.

4.3.1 Muon Lifetime Measurement

One of the most precise ingredients of the electroweak data set is the Fermi constant G_F, which is determined in measurements of the muon lifetime, τ_μ. The expression for the inverse lifetime

$$\frac{1}{\tau_\mu} = \frac{G_\mathrm{F}^2 m_\mu^5}{192\pi^3}(1 + \Delta q) \tag{4.12}$$

includes phase space, QED and hadronic corrections to the lowest order formula, summarised into Δq. This term is calculated to second order QED [14] so that the relative theoretical uncertainty on G_F is less than 0.3 ppm. The two recent experiments by the MuLan [15] and FAST [16] collaborations analysed the decay time of muons that are brought to rest in a target. Both experiments measured the decay rate of positrons from the reaction $\mu^+ \to e^+ \nu_e \bar{\nu}_\mu$ in scintillator detectors. MuLan is using a pulsed muon beam and scintillator detectors that are arranged spherically around the target, while FAST works with a continuous muon beam, and the scintillator pixel detector is the actual target. After background subtraction and noise correction, the muon lifetime is extracted from the exponential decay time spectrum with the following results [15, 16]

$$\tau_\mu\ (\mathrm{FAST}) = 2.197083(32)(15)\,\mu\mathrm{s} \tag{4.13}$$

$$\tau_\mu\ (\mathrm{MuLan}) = 2.197013(21)(11)\,\mu\mathrm{s}\ , \tag{4.14}$$

where the first error is statistical and the second systematic. The corresponding values of G_F improve on the current world average (WA) [17]:

$$G_\mathrm{F}\ (\mathrm{MuLan}) = 1.166371(6) \times 10^{-5}\ \mathrm{GeV}^{-2} \tag{4.15}$$

$$G_\mathrm{F}\ (\mathrm{FAST}) = 1.166353(9) \times 10^{-5}\ \mathrm{GeV}^{-2} \tag{4.16}$$

$$G_\mathrm{F}\ (\mathrm{WA}) = 1.16637(1) \times 10^{-5}\ \mathrm{GeV}^{-2}\ . \tag{4.17}$$

For the current analysis of electroweak data this improvement is however not so relevant because the precision of G_F largely exceeds those of the other electroweak parameters. The last value of G_F is therefore currently used in combined data analyses. Still, it is nice to see that the more recent measurements are in good agreement with this value.

4.3.2 Atomic Parity Violation

Sensitivity to electroweak parameters is also given in atomic parity violation experiments. Z boson contributions in the interaction of electrons and nucleus lead to parity violation when atomic hyperfine amplitudes are studied. The currently most precise experiments [18] studied the parity violating, highly forbidden 6S–7S transition in Caesium, $^{133}_{55}$Cs. The experiments shine two polarised laser beams on Cs vapour to first pump the atoms into the excited state and then use the second beam as analyser. An external electric field is applied to create an additional Starck-induced component which interferes with the parity violating transition. The asymmetry between two perpendicular polarisation directions of the analysing laser beam is then proportional to the parity violation strength.

The corresponding probability amplitude is calculated precisely in atomic many-body theory. It is proportional to the weak charge of the atomic nucleus

$$Q_W(Z,N) = -2\left\{C_{1u}(2Z+N) + C_{1d}(Z+2N)\right\} \ , \tag{4.18}$$

for an atom with Z protons and N neutrons. The weak charges of the up and down quark, C_{1u} and C_{1d}, are proportional to the axial and vector couplings of Eq. (1.28):

$$C_{1q} = 2g_{Ae}g_{Vq} \ . \tag{4.19}$$

The atomic weak charge can therefore also be written as

$$Q_W(Z,N) \approx Z(1-4\sin^2\theta_{\rm w}) - N \tag{4.20}$$

The most recent calculation [19] yields $Q_W(Cs) = -72.74 \pm 0.46$.

4.3.3 *Møller Scattering at EA316*

The Møller scattering experiment EA316 at the End Station A [20] at SLAC used a high-intensity pulsed and polarised electron beam of 45 and 48 GeV energy which passed through a cylinder filled with liquid hydrogen of 1.57 m length. The scattered electrons were detected in a magnetic spectrometer in an angular range $4.4 < \theta_{\rm lab} < 7.5$ mrad in the laboratory frame. Møller scattering electrons of 13–24 GeV are measured and separated from background of ep scattering in a radially segmented calorimeter. The average beam polarisation is $P_b = 0.89 \pm 0.04$.

In presence of neutral charge currents a parity violating asymmetry in the measured electron rate is expected between right-handed and left-handed incident beam:

$$A_{PV} = \frac{\sigma_R - \sigma_L}{\sigma_R + \sigma_L} \ . \tag{4.21}$$

After correcting for the polarisation P_b, calorimeter linearity, beam characteristics and backgrounds, the result obtained is

$$A_{PV} = -131 \pm 14\,(\text{stat.}) \pm 10\,(\text{syst.}) \times 10^{-9} \tag{4.22}$$

The relation to the effective weak mixing angle is given by

$$A_{PV} = -A(Q^2, y)\rho^{(e;e)}\left\{1 - 4\sin^2\theta_{\text{eff}}(Q) + \Delta\right\}, \tag{4.23}$$

where $\rho^{(e;e)}$ is the low-energy ratio of the weak neutral charge and the charged couplings, and Δ contains residual $O(\alpha)$ corrections. The effective analysing power is given by

$$A(Q^2, y) = \frac{G_F Q^2}{\sqrt{2}\pi\alpha_{\text{QED}}(Q)} \frac{1-y}{1+y^4+(1-y)^4} F_{\text{QED}} \tag{4.24}$$

where the momentum transfer is $Q^2 = 0.026\ \text{GeV}^2$, $y = Q^2/s \approx 0.6$. The factor $F_{\text{QED}} = 1.01 \pm 0.01$ takes QED radiative corrections into account, like ISR/FSR, box and vertex corrections. Numerically, A is determined from simulations to be $A = (3.25 \pm 0.05) \times 10^{-9}$. The electroweak mixing angle derived from this measurement is

$$\sin^2\theta_{\text{eff}}(Q^2) = 0.2397 \pm 0.0010\ (\text{stat.}) \pm 0.0008\ (\text{syst.}) \tag{4.25}$$

at the experimental $Q^2 = 0.026\ \text{GeV}^2$, in agreement with a Standard Model expectation of $\sin^2\theta_{\text{eff}}(Q^2) = 0.2381 \pm 0.0006$. After evolution to M_Z one obtains

$$\sin^2\theta_{\text{w}}\left(M_Z^2\right) = 0.2330 \pm 0.0011\ (\text{stat.}) \pm 0.0009\ (\text{syst.}) \pm 0.0006\ (\text{evolution}) \tag{4.26}$$

which is used as input to the electroweak data analysis.

4.3.4 Neutrino-Nucleon Scattering at NuTeV

A measurement of the weak mixing angle is also performed in ν and $\bar{\nu}$ nucleon interactions in the NuTeV detector [21], which was built as a 690 ton steel-scintillator target. The very pure muon neutrino and anti-neutrino beams undergo charged (CC) and neutral current (NC) interactions. Both NC and CC reactions create short-range hadronic cascades in the detector, only the CC reaction however produces a final state muon that penetrates the detector over a longer distance. The length of the events measured from the interaction vertex in units of traversed scintillator counters is therefore a characterisation of their NC and CC nature.

By counting those events the NC/CC ratio of the cross-section differences of ν and $\bar{\nu}$ reactions can be determined:

$$R^- = \frac{\sigma(\nu_\mu N \to \nu_\mu X) - \sigma(\bar{\nu}_\mu N \to \bar{\nu}_\mu X)}{\sigma(\nu_\mu N \to \mu^- X) - \sigma(\bar{\nu}_\mu N \to \mu^+ X)} \tag{4.27}$$

According to Paschos-Wolfenstein [22] this ratio is related to the left and right handed couplings of the neutrinos and the u and d valence quarks in the target by

$$R^- = 4g_{L\nu}^2 \sum_{u,d} \left\{g_{Lq}^2 - g_{Rq}^2\right\} = \rho_\nu \rho_{ud} \left\{\frac{1}{2} - \sin^2\theta_{\mathrm{w}}\right\} , \tag{4.28}$$

from which the on-shell value of $\sin^2\theta_{\mathrm{w}}$ can be extracted. NuTeV determines [21]

$$\begin{aligned}\sin^2\theta_{\mathrm{w}} &= 1 - \frac{M_{\mathrm{W}}^2}{M_{\mathrm{Z}}^2} \\ &= 0.22773 \pm 0.00135\,(\mathrm{stat.}) \pm 0.00093\,(\mathrm{syst.}) \\ &\quad -0.00022 \left(\frac{m_{\mathrm{t}}^2 - (175\ \mathrm{GeV})^2}{(50\ \mathrm{GeV})^2}\right) + 0.00032 \ln\left(\frac{M_{\mathrm{H}}}{150\ \mathrm{GeV}}\right)\end{aligned} \tag{4.29}$$

with the given dependence on m_{t} and M_{H}. This measured value disagrees from the Standard Model prediction by about 3σ. Figure 4.5 compares the low-Q^2 measurements to the Standard Model prediction, after conversion to $\sin^2\theta_{\mathrm{eff}}$. However, R^- is a derived quantity obtained from the individual NC/CC ratios for neutrinos and anti-neutrinos, R^ν and $R^{\bar{\nu}}$, which are measured by NuTeV directly. The deviation can be isolated to be mainly in R^ν which is also much more sensitive to $\sin^2\theta_{\mathrm{w}}$. The ratios R^ν and $R^{\bar{\nu}}$ are displayed separately in Fig. 4.5.

A complication of the NuTeV measurement arises however from the fact that also strange sea-quark contributions need to be taken into account in the parameter extraction. A detailed analysis of the strange/anti-strange asymmetry component in the PDFs yields a negative, but small correction of -0.0014 ± 0.0010 [17, 24] to $\sin^2\theta_{\mathrm{w}}$, reducing the Standard Model difference to $1.9\,\sigma$. Furthermore, new evaluations of the radiative correction terms involving QED hard photon emission [25]

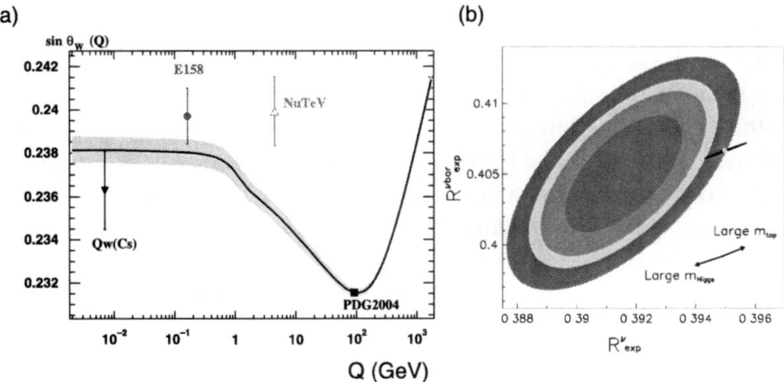

Fig. 4.5 (**a**) Evolution of the electroweak mixing angle $\sin^2\theta_{\mathrm{eff}}$ as a function of momentum transfer Q, compared to the measurements in atomic parity violation, Møller scattering, neutrino-nucleon scattering and at the Z pole [20]. (**b**) NC/CC ratios as measured by NuTeV separately in neutrino and anti-neutrino nucleon scattering, R^ν and $R^{\bar{\nu}}$ [23]. The anti-neutrino ratio $R^{\bar{\nu}}$ is in agreement with the Standard Model prediction, shown as a point, while R^ν is not. In a recent re-analysis of the data [17] the difference of the to the Standard Model prediction is reduced to 1.9 standard deviations

may lead to additional corrections of the measurement value and proper treatment of PDFs is needed [17, 26]. Scrutinising the NuTeV result is therefore still ongoing. New measurements with higher resolution are proposed [27] and may resolve the situation experimentally.

4.3.5 Running of α_{QED}

The running of the electromagnetic coupling α_{QED} is a consequence of the screening of the electromagnetic charge by polarisation of the vacuum. The running is experimentally established in the measurement of Bhabha scattering $e^+e^- \to e^+e^-$ at LEP, where the differential cross-section $d\sigma/dt$ is determined. The dependence on the electromagnetic coupling is [28]:

$$\frac{d\sigma}{dt} = \frac{4\pi\alpha_{\text{QED}}(t^2)}{t^2}(1+\varepsilon)(1+\delta_\gamma)(1+\delta_z) \,, \tag{4.30}$$

where the s-channel contributions δ_γ and δ_Z are much smaller than the radiative corrections ε. The overall scale of the cross-section is however not a good measure, because it is normalised by the beam luminosity which is derived from low-angle Bhabha scattering which again assumes $\alpha_{\text{QED}}(t^2)$ to be known. On the other hand the angular distribution contains information on $\alpha_{\text{QED}}(t^2)$ since

$$Q^2 = t \approx -s\frac{1-\cos\theta}{2} \,, \tag{4.31}$$

and s is known precisely at LEP. Such a measurement is performed by OPAL at low Q^2 [28] in the very forward luminosity monitor with its very high angular resolution and by L3 in both low and high Q^2 regions [29]. Figure 4.6 summarises the LEP results [30] and shows clearly that data are incompatible with a constant value of α_{QED} and that the running follows the QED prediction.

For the electroweak data analysis a more precise method is however necessary. As mentioned in Chap. 1, the leptonic and top quark part of $\Delta\alpha(M_Z^2)$ are calculable with small theoretical uncertainty. Experimental input is then used to derive the hadronic contributions, $\Delta\alpha^{(5)}_{\text{had}}$, applying the dispersion relation [31]:

$$\Delta\alpha^{(5)}_{\text{had}}(Q^2) = -\frac{Q^2\alpha_{\text{QED}}}{3\pi}\int_0^\infty ds\frac{R(s)}{s(s-Q^2)} \tag{4.32}$$

with the hadronic to leptonic e^+e^- annihilation cross-section ratio

$$R(s) = \frac{\sigma^0(e^+e^- \to \text{hadrons})}{\sigma^0(e^+e^- \to \mu^+\mu^-)} \,, \text{and } \sigma^0(e^+e^- \to \mu^+\mu^-) = \frac{4\pi\alpha^2_{\text{QED}}}{3s} \,. \tag{4.33}$$

The ratio $R(s)$ is calculable in perturbative QCD only in regions away from $q\bar{q}$ thresholds and resonances. The approach of [32] is therefore to apply theoretical

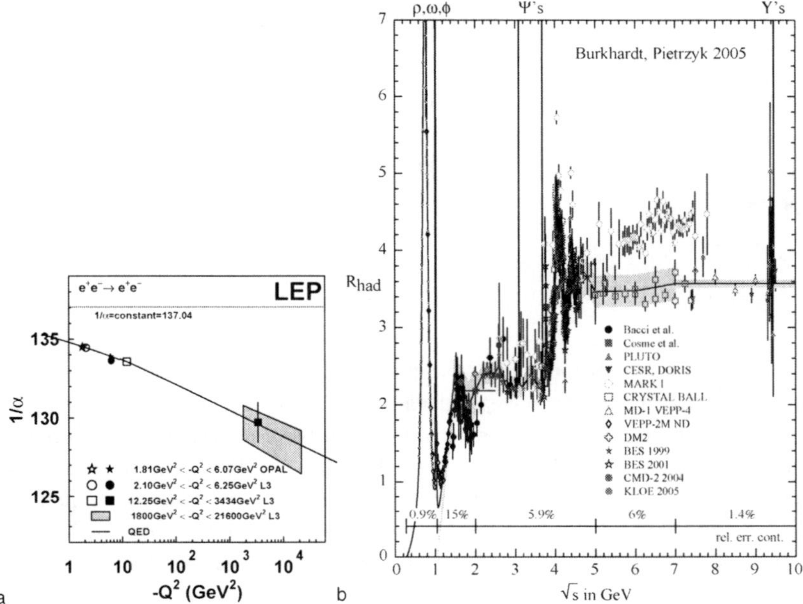

Fig. 4.6 (**a**) Electromagnetic coupling constant measured by LEP in Bhabha scattering as a function of Q^2 compared to a constant value of $\alpha_{\rm QED}$ and the QED prediction [30]. (**b**) Data and corresponding uncertainty used to determine $R(s)$ in regions where perturbative QCD can not be safely applied [32]

calculations for $\sqrt{s} > 12$ GeV and experimental data for lower $\sqrt{s}$, which is shown in Fig. 4.6. This yields the already mentioned value of [32]:

$$\Delta\alpha_{\rm had}^{(5)} = 0.02758 \pm 0.00035\ . \tag{4.34}$$

A more theoretically driven value is obtained in [31], where perturbative QCD is used to calculate $R(s)$ also for $\sqrt{s} > 1.4$ GeV, with the result

$$\Delta\alpha_{\rm had}^{(5)} = 0.02749 \pm 0.00012\ . \tag{4.35}$$

The two estimations agree well, and also with other independent determinations [17].

4.3.6 Anomalous Magnetic Moment of the Muon

Charged elementary particles with half-integer spin have a magnetic dipole moment μ parallel to the spin **s**:

$$\mu = g\frac{q}{2m}\mathbf{s}\ . \tag{4.36}$$

with the Landé g factor. For leptons, g has a value of about 2, and the exact value depends on radiative corrections, summarised in the anomaly:

$$a_\ell = \frac{g-2}{2}\,. \tag{4.37}$$

The measurements of these parameters are sensitive to higher order quantum corrections from QED, electroweak theory, hadronic contributions and possibly new physics beyond the Standard Model. The sensitivity of a_ℓ to high mass scales Λ is proportional to

$$\frac{\delta a_\ell}{a_\ell} = \frac{m_\ell^2}{\Lambda^2} \tag{4.38}$$

so that it is more advantageous to measure $g-2$ with muons than with electrons. The most precisely measured value of a_μ is obtained by the E821 experiment [33]. Polarised muons from a pion beam were stored in a superconducting magnet ring producing a highly uniform **B** field of 1.45 T. The degree of polarisation reached 95%. For vertical focusing of the muon beam, electric quadrupoles were arranged around the ring which had a central orbital radius of 7.11 m. The cyclotron frequency taking both electrical and magnetic fields into account is given by [34]

$$\boldsymbol{\omega}_C = -\frac{q}{m}\left\{\frac{\mathbf{B}}{\gamma} - \frac{\gamma}{\gamma^2-1}\boldsymbol{\beta}\times\mathbf{B}\right\}, \tag{4.39}$$

and the spin precession frequency is equal to

$$\boldsymbol{\omega}_S = -\frac{q}{m}\left\{\left(\frac{g-2}{2}+\frac{1}{\gamma}\right)\mathbf{B} - \frac{g-2}{2}\frac{\gamma}{\gamma+1}(\boldsymbol{\beta}\cdot\mathbf{B})\boldsymbol{\beta} - \left(\frac{g}{2}-\frac{\gamma}{\gamma+1}\right)\boldsymbol{\beta}\times\mathbf{E}\right\}, \tag{4.40}$$

with the relativistic γ factor and the velocity $\boldsymbol{\beta}$ of the muons. For a magnetic field perpendicular to the muon momentum, $\beta\cdot\mathbf{B}=0$, the spin precession relative to the momentum occurs at a frequency

$$\begin{aligned}\boldsymbol{\omega}_a &= \boldsymbol{\omega}_S - \boldsymbol{\omega}_C \\ &= -\frac{q}{m}\left\{a_\mu\mathbf{B} - \left(a_\mu - \frac{1}{\gamma^2-1}\right)\boldsymbol{\beta}\times\mathbf{E}\right\}.\end{aligned} \tag{4.41}$$

The dependence on the electric field is eliminated for the "magic" $\gamma = 29.3$ [35], which corresponds to a muon momentum of $p = 3.09$ GeV. Due to this simplification, a_μ can be determined from a measurement of ω_a and B. The latter is measured with proton-NMR based on water probes, using the proton Larmor frequency

$$\omega_p = g_p\frac{eB}{2m_p}\,. \tag{4.42}$$

The muon anomalous magnetic moment is therefore derived from the relation

$$a_\mu = \frac{\omega_a/\omega_p}{\lambda - (\omega_a/\omega_p)} \tag{4.43}$$

with $\lambda = \mu_\mu/\mu_p = 3.18334539(10)$ [36]. The value of ω_p is calibrated to a spherical H_2O probe with a small systematic uncertainty of 0.17×10^{-9}.

What remains to be measured is ω_a, which is done by detecting the decay electron rate of the parity violating muon decay $\mu \to e\nu_e\nu_\mu$. The preferred direction of the decay electron with respect to the muon spin depends on the electron energy [34]. Applying a calorimetric selection of electrons above a certain energy threshold, E_{th}, the measured electron rate follows the time dependence

$$N(t, E_{\text{th}}) = N_0(E_{\text{th}})e^{-t/\gamma\tau_\mu}\,[1 + A(E_{\text{th}})\cos(\omega_a t + \phi(E_{\text{th}})]\ , \tag{4.44}$$

with the asymmetry parameter

$$A(E_{\text{th}}) = P\frac{y_{\text{th}}(2y_{\text{th}} + 1)}{-y_{\text{th}}^2 + y_{\text{th}} + 3}\ . \tag{4.45}$$

This parameter depends on the polarisation $P \approx 95\%$ of the muon beam and the energy ratio $y_{\text{th}} = E_{\text{th}}/E_{\text{max}}$, where the maximal electron energy in the laboratory frame is $E_{\text{max}} = 3.09$ GeV. Since the momenta of the decay electrons are smaller than the muon momenta, the electrons are swept to the inside of the storage ring and are detected in 24 scintillating fibre calorimeters, evenly placed around the ring. A typical energy threshold is 1.8 GeV and the corresponding asymmetry, A, is about 35%. Figure 4.7 shows the time dependence of the detected number of electrons with the typical oscillation structure overlaid on an exponential decay curve.

The detailed data analysis takes the beam dynamics, electrical field corrections, pitch effects, magnetic field systematics into account. The final experimental value is

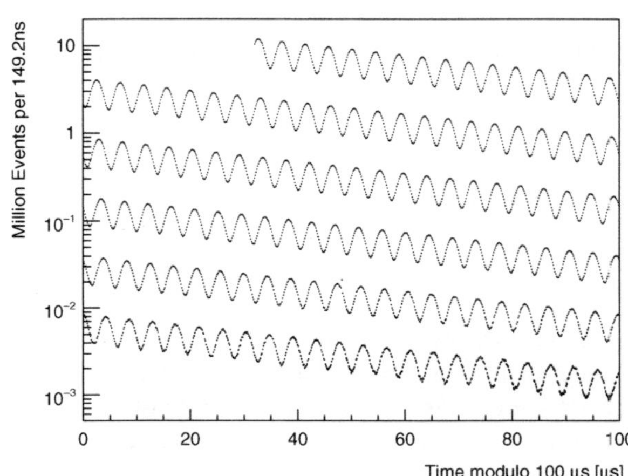

Fig. 4.7 **(a)** Total number of electron above 1.8 GeV as a function of time (modulo 100 μs) from the 2001 μ^- data set of E821 (from [34])

$$a_\mu(\text{exp.}) = 11659208.0(6.3) \times 10^{-10} , \tag{4.46}$$

with a statistical uncertainty of 5×10^{-10} and a systematic uncertainty of 4×10^{-10}. The total relative uncertainty is 5×10^{-10}.

This high experimental precision is a great challenge for the theoretical calculation of the expected value of a_μ. The first order QED result [37]

$$a_\mu = \frac{\alpha_{\text{QED}}}{2\pi} \tag{4.47}$$

is known since a long time already. The recent calculations cover:

- complete QED 4-loop results, 5-loop leading logarithmic corrections, and an estimation of the remaining 5-loop corrections, which would involve 12672 Feynman diagrams but is beyond the accuracy needed [34, 38]

$$a_\mu(\text{QED}) = (1165847180.9 \pm 1.4_{5-\text{loops}} \pm 0.8_{\alpha_{\text{QED}}} \pm 0.4_{\text{masses}}) \times 10^{-10} \tag{4.48}$$

- hadronic vacuum polarisation [39]

$$a_\mu(\text{had.}) = (690.8 \pm 4.4) \times 10^{-10} \tag{4.49}$$

- hadronic light-by-light scattering [40, 41]

$$a_\mu(\text{lbl.}) = (+10.5 \pm 2.6) \times 10^{-10} \tag{4.50}$$

- higher order hadronic vacuum polarisation [40]

$$a_\mu(\text{had.,ho.}) = (-9.8 \pm 0.1) \times 10^{-10} \tag{4.51}$$

- electroweak contribution in leading 2-loop and 3-loop order [42, 40]

$$a_\mu(\text{EW}) = (15.2 \pm 0.2) \times 10^{-10}, \tag{4.52}$$

The result of these complex calculations is the Standard Model prediction of [40]

$$a_\mu(\text{SM}) = (11659178.5 \pm 5.1) \times 10^{-10} . \tag{4.53}$$

The deviation of the theory from experiment

$$\Delta a_\mu = a_\mu(\text{exp.}) - a_\mu(\text{SM}) = (29.5 \pm 8.1) \times 10^{-10} \tag{4.54}$$

is at the level of 3.6 standard deviations. The theoretical uncertainty is slightly smaller than the experimental one and all recent theoretical approaches [17, 34, 40] are showing a discrepancy in the order of 3 standard deviations or more. Some uncertainty is however still in the treatment of the hadronic contributions to a_μ(SM), which are usually derived from $e^+e^- \to$ hadrons data. If τ-spectral functions corrected for isospin-breaking effects are used instead, the deviation from the Standard Model is only about 2 sigma [43]. Preliminary data from $\pi^+\pi^- + \gamma_{ISR}$ production with initial state photons taken at the $\Upsilon(4S)$ resonance by the BaBar collaboration point in the same direction [44]. More experimental and theoretical understanding of the different ways to extract the hadronic corrections to a_μ(SM) is thus necessary.

In general it is however interesting to observe that the electroweak corrections are small compared to the hadronic uncertainties, so that they cast dependencies on the Standard Model parameters, including $M_{\rm H}$. The electroweak term can be written as [34, 42]

$$\begin{aligned} a_\mu(\mathrm{EW}) &= \frac{G_{\rm F}}{\sqrt{2}}\frac{m_\mu^2}{8\pi^2}\left\{\frac{5}{3}+\frac{1}{3}\left(1-4\sin^2\theta_{\rm w}\right)^2-\frac{\alpha_{\rm QED}}{\pi}[155.5(4)(1.8)]\right\} \\ &= (15.4\pm0.2\pm0.1)\times10^{-10}\ . \end{aligned} \tag{4.55}$$

The a_μ dependence on the mass of the Standard Model Higgs boson is only of the order $\frac{G_{\rm F}}{\sqrt{2}}\frac{m_\mu^2}{4\pi}\frac{m_\mu^2}{M_{\rm H}^2}\log\frac{M_{\rm H}^2}{m_\mu^2}$ [34], and determines the first $\pm0.2\times10^{-10}$ error, while the second is from higher-order hadronic effects.

On the other hand, new physics may enter the game through loop contributions. Super-symmetric particles, for example, would give rise to an additional term [45]

$$a_\mu(\mathrm{SUSY}) \approx 13\times10^{-10}\left(\frac{100\ \mathrm{GeV}}{M_{\rm SUSY}}\right)^2\tan\beta\,\mathrm{sign}(\mu)\ , \tag{4.56}$$

where $M_{\rm SUSY}$ denotes the common mass scale of SUSY particles, $\tan\beta$ the ratio of the two Higgs vacuum expectation values of the two Higgs doublets, and μ the Higgsino mass parameter. Assuming SUSY masses in the order of 200 GeV, i.e. in the near reach of the LHC, a value of $\tan\beta\approx 6$ would compensate the full $\approx 3\,\sigma$ difference between experiment and theoretical prediction.

4.4 Model Analysis of Electroweak Data

The data set of precision electroweak measurements consists of

- precise $M_{\rm Z}$ and $M_{\rm W}$ measurements at LEP, SLD, and Tevatron,
- precise Z line-shape data from LEP and SLD, including Z width, branching fractions, and decay asymmetries,
- the top mass determination at the Tevatron,
- measurements at low Q^2.

The Standard Model predictions for the electroweak observables are calculated using the TOPAZ0 [46] and ZFITTER [47] programs. They contain higher order QCD corrections, two-loop corrections for $M_{\rm W}$, complete fermionic two-loop corrections for $\sin^2\theta^\ell_{\rm eff}$, and three-loop top-quark contributions to the ρ parameter [13].

The theoretical uncertainties on M_{W} are 4 MeV and 0.000049 on $\sin^2\theta^{\ell}_{\mathrm{eff}}$. As stated in [13], the latter dominates the theoretical uncertainty in Standard Model fits and the indirect M_{H} extraction, which could be improved by a more accurate, two-loop calculation of the partial Z decay widths.

The measured high-Q^2 observables are compared to the theoretical predictions in Fig. 4.8. The free parameters in the fit to data are $\Delta\alpha^{(5)}_{\mathrm{had}}$, $\alpha_s(M_Z^2)$, M_{Z}, m_{t}, M_{H}. The agreement is good, and expressed in terms of χ^2/DoF a value of 17.3/13 is obtained, which corresponds to a fit probability of 18%. Adding the low-Q^2 measurements, excluding a_μ, one gets a $\chi^2/\mathrm{DoF} = 27.5/17$ and a probability of only 5.2%. This is mainly due to the NuTeV neutrino-nucleon scattering result, as discussed above. The muon anomalous magnetic moment has only negligible sensitivity to electroweak parameters, but should still be added to the global χ^2. Doing this, one finally obtains a χ^2/DoF of 40.5/18 or a probability of 0.2%. The "outliers" are easily identified:

- the forward-backward asymmetry of b quarks at the Z pole, $A^{0,b}_{\mathrm{FB}}$, with 2.9 σ, and
- the muon anomalous magnetic moment, a_μ, with 3.6 σ.

Both gave rise to numerous theoretical speculations and explanations. An attractive one exists for the latter, a_μ. In the Minimal Super-symmetric Standard Model (MSSM) additional contributions from elementary super-partner particles in the first and second order loops lead to additional radiative corrections that shift the expectation value closer to the measurement [45].

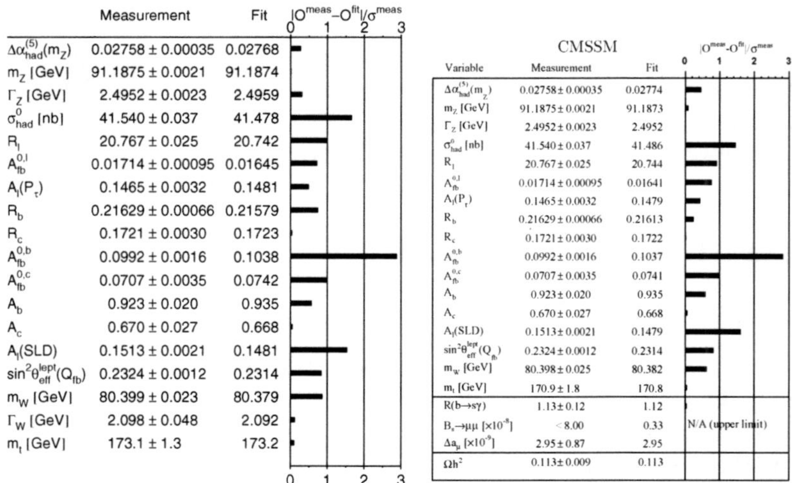

	Measurement	Fit
$\Delta\alpha^{(5)}_{had}(m_Z)$	0.02758 ± 0.00035	0.02768
m_Z [GeV]	91.1875 ± 0.0021	91.1874
Γ_Z [GeV]	2.4952 ± 0.0023	2.4959
σ^0_{had} [nb]	41.540 ± 0.037	41.478
R_l	20.767 ± 0.025	20.742
$A^{0,l}_{fb}$	0.01714 ± 0.00095	0.01645
$A_l(P_\tau)$	0.1465 ± 0.0032	0.1481
R_b	0.21629 ± 0.00066	0.21579
R_c	0.1721 ± 0.0030	0.1723
$A^{0,b}_{fb}$	0.0992 ± 0.0016	0.1038
$A^{0,c}_{fb}$	0.0707 ± 0.0035	0.0742
A_b	0.923 ± 0.020	0.935
A_c	0.670 ± 0.027	0.668
A_l(SLD)	0.1513 ± 0.0021	0.1481
$\sin^2\theta^{lept}_{eff}(Q_{fb})$	0.2324 ± 0.0012	0.2314
m_W [GeV]	80.399 ± 0.023	80.379
Γ_W [GeV]	2.098 ± 0.048	2.092
m_t [GeV]	173.1 ± 1.3	173.2

CMSSM

Variable	Measurement	Fit	$\lvert O^{meas}-O^{fit}\rvert/\sigma^{meas}$
$\Delta\alpha^{(5)}_{had}(m_Z)$	0.02758±0.00035	0.02774	
m_Z [GeV]	91.1875±0.0021	91.1873	
Γ_Z [GeV]	2.4952±0.0023	2.4952	
σ^0_{had} [nb]	41.540±0.037	41.486	
R_l	20.767±0.025	20.744	
$A^{0,l}_{fb}$	0.01714±0.00095	0.01641	
$A_l(P_\tau)$	0.1465±0.0032	0.1479	
R_b	0.21629±0.00066	0.21613	
R_c	0.1721±0.0030	0.1722	
$A^{0,b}_{fb}$	0.0992±0.0016	0.1037	
$A^{0,c}_{fb}$	0.0707±0.0035	0.0741	
A_b	0.923±0.020	0.935	
A_c	0.670±0.027	0.668	
A_l(SLD)	0.1513±0.0021	0.1479	
$\sin^2\theta^{lept}_{eff}(Q_{fb})$	0.2324±0.0012	0.2314	
m_W [GeV]	80.398±0.025	80.382	
m_t [GeV]	170.9±1.8	170.8	
$R(b\to s\gamma)$	1.13±0.12	1.12	
$B_s\to\mu\mu$ [$\times10^{-8}$]	<8.00	0.33	N/A (upper limit)
Δa_μ [$\times10^{-9}$]	2.95±0.87	2.95	
Ωh^2	0.113±0.009	0.113	

Fig. 4.8 Compatibility of electroweak data with the SM [13] (*left*) and the CMSSM [48] (*right*). The *left table* compares most recent data, including 2009 results, while the *right table* is recent, but not updated with latest results. The level of agreement with electroweak data is however practically unchanged

A global data analysis in the constrained MSSM (CMSSM) was performed in [48], with the result shown in Fig. 4.8. Apart from a_μ, the SUSY-sensitive heavy flavour observables $Br(b \to s\gamma)$, $Br(B_s \to \mu^+\mu^-)$, and the cold dark matter (CDM) density in the universe, $\Omega_c h^2$.

The first two are rare b-decay processes that are induced by penguin loops in the Standard Model. They get enhanced by additional SUSY loop-contributions which are proportional to $\tan\beta$ and $\tan^6\beta$ [45], respectively. The most recent measurements [49] are $Br(b \to s\gamma) = (352 \pm 23 \pm 9) \times 10^{-6}$, and $Br(B_s \to \mu^+\mu^-) < 2.3 \times 10^{-8}$. The CMSSM analysis uses the experimental to theory ratio $R(b \to s\gamma) = Br(b \to s\gamma)/Br_{SM}(b \to s\gamma)$ in the fit.

The dark matter density is measured from a scan of the cosmic microwave background (CMB) by WMAP [50]. A fit to the angular power spectrum of temperature and polarisation data, adding further information from small and large scale cosmic structures, yields a matter density of $\Omega_m h^2 = 0.128 \pm 0.008$ and a baryonic density of $\Omega_b h^2 = 0.0223 \pm 0.007$, so that the cold dark matter density is $\Omega_c h^2 = \Omega_m h^2 - \Omega_b h^2 = 0.106 \pm 0.009$. Assuming super-symmetry with R-parity conservation, i.e. with a Lightest Super-symmetric Particle (LSP) that can not decay, this particle contributes to the CDM density. The predicted value of $\Omega_c h^2$ actually is proportional to the mass of the LSP and to the thermally averaged annihilation cross-section of the LSP to Standard Model particles. Theoretical models (e.g. [51]) take details of the annihilation and co-annihilation processes into account in the determination of the temperature dependent abundances of dark matter.

The CMSSM (or mSUGRA) analysis [48] assumes the soft SUSY-breaking scalar masses to be universal at the GUT scale, with a value of M_0, as well as the soft SUSY-breaking gaugino masses, $M_{1/2}$, and the trilinear couplings, A_0. Furthermore, the ratio of the two vacuum expectation values of the Higgs doublets, $\tan\beta$, and the sign of the Higgs mixing parameter μ are varied. The parameters at the electroweak scale are derived from renormalisation group equations (RGE). As can be seen in Fig. 4.8, the CMSSM fit describes the observables used in the previous analysis as well as the Standard Model fit. The constraints from heavy flavour decays and the CDM density are also fulfilled and in agreement with the measured values of $R(b \to s\gamma)$, $Br(B_s \to \mu^+\mu^-)$ and $\Omega_c h^2$. The main difference to the Standard Model analysis is the much better agreement with the observed value of a_μ. In terms of χ^2 the CMSSM fit reaches of 17.34 per 14 of freedom, which corresponds to a fit probability of 24%. The CMSSM is therefore removing parts of the discrepancy between theory and experiment, but can not explain all deviations, like the 2.7 σ effect in $A_{\rm FB}^{0,\rm b}$.

Another comparison of Standard Model and MSSM predictions [52] with data is shown in Fig. 4.9, where the theory is confronted with the current W and top mass measurements. As can be observed, the Standard Model is in agreement with these data if the Higgs boson mass is small and close to the LEP exclusion limit of $M_{\rm H} > 114.4$ GeV [53]. On the other hand, SUSY is fitting well if the SUSY mass scale is not too light.

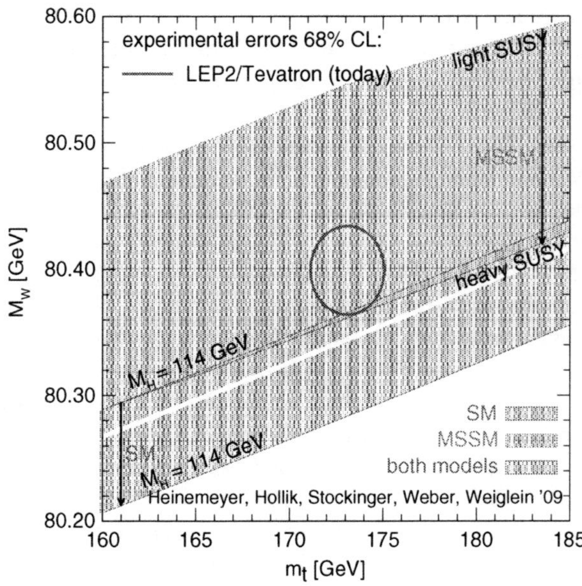

Fig. 4.9 Direct measurement of W and top mass, compared to Standard Model calculations with varying Higgs mass and to MSSM predictions with different SUSY mass scales [52]

4.5 Electroweak Constraints on New Particles

The last missing particle to complete the Standard Model is the Higgs boson, which up to now has not been discovered yet. Theoretical arguments (see Chap. 1) indicate that its mass should be in the sub-TeV region. The currently most stringent search limits are from the LEP and Tevatron experiments. The final LEP result is shown in Fig. 4.10, from which a lower limit of $M_{\mathrm{H}} > 114.4$ GeV at 95% C.L. [53] is derived. Very recent search results by CDF and DØ using up to 4.2 fb^{-1} of $p\bar{p}$ data per experiment, shown as well in Fig. 4.10, furthermore exclude a Higgs mass between 160 and 170 GeV at 95% C.L. [54]. With the full data set of projected 2×10 fb^{-1} until the end of 2011, a larger mass interval will be covered.

However, as long as the Higgs boson or super-symmetric particles are not discovered, the analysis of current electroweak data can give hints about the possible mass range of these particles.[1] Figure 4.11 shows the sensitivity of some of the electroweak parameters to the Higgs mass in terms of the corresponding constraints on M_{H} that can be derived. All data combined are used in the Standard Model analysis which yields the left $\Delta\chi^2$ curve of Fig. 4.11. Theoretical uncertainties are indicated by the band, but they do not change the general behaviour of a clear minimum at

[1] A similar chain of events happened already in case of the top quark discovery [55].

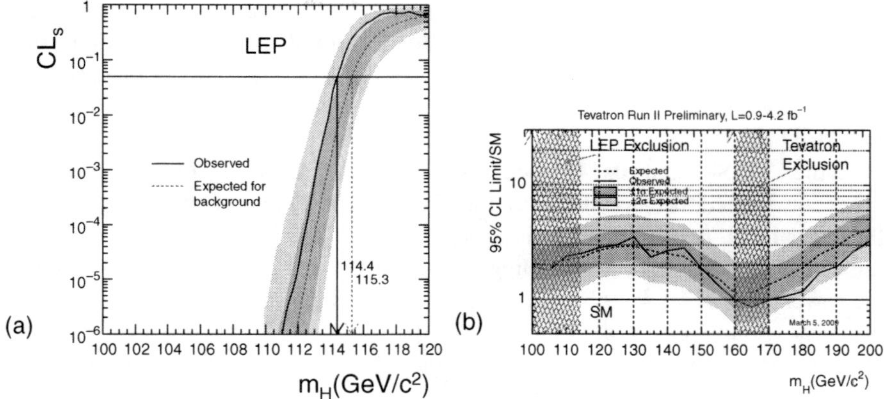

Fig. 4.10 (**a**) Confidence level ratio of the signal+background hypothesis, CL_s, as observed and expected in LEP data as a function of M_H [53]. The lower limit on M_H is derived from the crossing point with the $CL_s = 0.95$ line. The observed limit of 114.4 GeV is slightly *lower* than expected. The bands indicate the 65 and 95% probability regions of the expected limit in absense of any signal. (**b**) Observed and expected 95% C.L. upper limits on the ratio of the Standard Model cross-section of Higgs production, as a function of M_H for the combined, preliminary Tevatron data. The bands indicate the 65 and 95% probability regions of the expected limit in absense of any signal. The observed limit corresponds to an exclusion of a Higgs mass between 160 and 170 GeV at 95% C.L. [54]

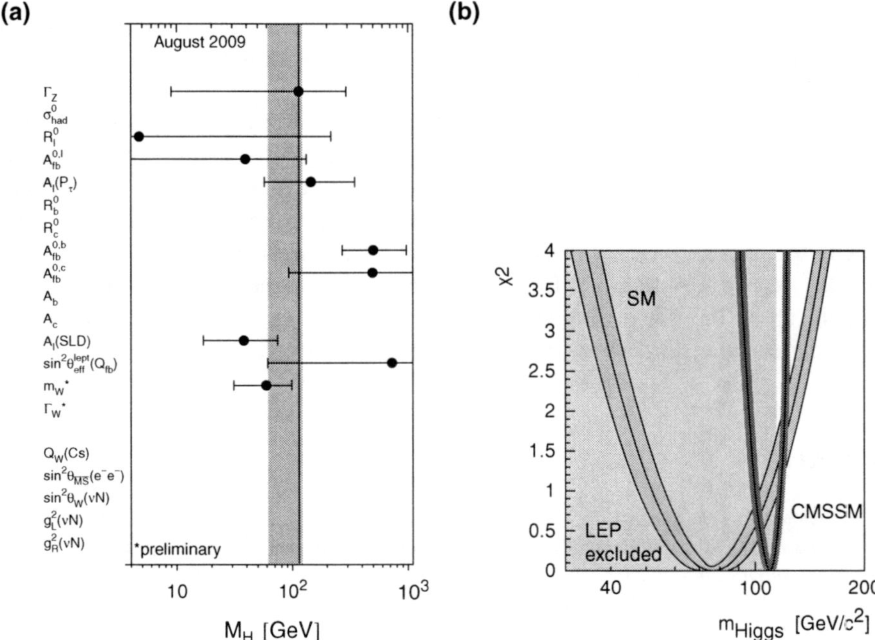

Fig. 4.11 (**a**) Electroweak measurements and their individual constraints on the Higgs mass [12]. (**b**) The sensitivity of the SM and CMSSM Higgs boson h expressed in terms of $\Delta\chi^2$ of the combined fit to the electroweak data [48]

rather low $M_{\rm H}$. Not taking into account the low-Q^2 data, the χ^2 minimum is at a Higgs mass of 90^{+36}_{-27} GeV. Since the dependence of radiative corrections on $M_{\rm H}$ is logarithmic (see Eq. 1.50), the one-sided 95% confidence level upper limit is found at 163 GeV, increasing to 191 GeV when the lower direct Higgs mass bound is included [13].

In the CMSSM analysis [48] the mass of the lightest CP-even Higgs boson is even more constrained, because there is a theoretical upper bound of about 135 GeV. On the other hand, the minimal χ^2 is at $M_h = 110^{+8}_{-10}$ (exp.) $\pm$ 3 (theo.) GeV, and higher than the Standard Model Higgs mass. This is in much better agreement with the lower bounds on M_h obtained by LEP for the lightest super-symmetric Higgs boson which are in the order of 90 GeV [56].

The CMSSM fit [48] gives further indications on preferred parameter values, which are best compatible with data:

$$\begin{aligned} M_0 &= 85^{+40}_{-28}\ \text{GeV} \\ M_{1/2} &= 280^{+140}_{-30}\ \text{GeV} \\ A_0 &= -360^{+300}_{-140}\ \text{GeV} \\ \tan\beta &= 10^{+9}_{-4}\ \text{GeV} \\ \text{sign}(\mu) &= +1\ \text{(fixed)}\ . \end{aligned}$$

With this parameter set, the super-symmetric particle mass spectrum consists of a light Higgs boson at $M_h \approx 100$ GeV, charged and heavy Higgs bosons at $M_{H^\pm}, M_H, M_A \approx 450$–$500$ GeV, and the lightest neutralino as LSP at $M_{\tilde{\chi}^0_1} \approx 100$–$120$ GeV.

The prospects for the Large Hadron Collider to discover the Standard Model Higgs boson are therefore excellent, although challenging in the low $M_{\rm H}$ region, as will be discussed in the Chap. 7. If the electroweak and Higgs sector of the Standard Model is not realised in its current form and if super-symmetry is the extension nature has chosen, the LHC is as well the ideal machine to push the discovery frontier far beyond Tevatron and LEP. Assuming, for example, the CMSSM scenario and the above described parameter set, only 1 fb^{-1} of pp data will be sufficient to see clear evidence for super-symmetric particle production [57].

References

1. Tevatron Electroweak Working Group, *Combination of CDF and D0 Results on the W Boson Mass and Width*, arXiv:0808.0147v1; Tevatron Electroweak Working Group, *Updated Combination of CDF and D0 Results for the Mass of the W Boson*, arXiv:0908.1374v1.
2. The CDF Collaboration, T. Aaltonen, et al., Phys. Rev. Lett. **99** (2007) 151801; Phys. Rev. **D 77** 112001.
3. J. Pumplin, D. R. Stump, J. Huston, H. L. Lai, P. Nadolsky and W. K. Tung, JHEP **0207** (2002) 012, hep-ph/0201195; J. Pumplin, et al., *Parton Distributions and the Strong Coupling Strength: CTEQ6AB PDFs*, hep-ph/0512167.

4. A. D. Martin, W. J. Stirling, R. S. Thorne and G. Watt, *Update of Parton Distributions at NNLO*, arXiv:0706.0459v3.
5. The DØ Collaboration, V. Abazov, et al., Phys. Rev. **D66 012001** (2002); The DØ Collaboration, V. Abazov, et al., *Measurement of the W Boson Mass*, arXiv:0908.0766.
6. M. Jezabek and J. H. Kühn, Nucl. Phys. **B 314** (1989) 1.
7. The CDF Collaboration, CDF note 8104, 2006.
8. The CDF Collaboration, CDF Conf. Note 9692; hep-ex/0812.4469v1.
9. The DØ Collaboration, DØ note 5877-CONF, 2009.
10. The CDF Collaboration, CDF Conf. Note 9694, 2009.
11. Tevatron Electroweak Working Group, for the CDF Collaboration, the DØ Collaboration, *Combination of CDF and D0 Results on the Mass of the Top Quark*, arXiv:0903.2503v1.
12. The CDF Collaboration, CDF note 9414, 2008.
13. The ALEPH, DELPHI, L3, OPAL, SLC Collaborations, the LEP Electroweak Working Group, the Tevatron Electroweak Working Group, and the SLD electroweak and heavy flavour groups, *Precision Electroweak Measurements and Constraints on the Standard Model*, CERN-PH-EP/2008-020; arXiv:0811.4682. Updates can be found at http://lepewwg.web.cern.ch/LEPEWWG/.
14. T. van Ritbergen and R. G. Stuart, Nucl. Phys. **B 564** (2000) 343; T. van Ritbergen and R. G. Stuart, Phys. Lett. **B 437** (1998) 201; T. van Ritbergen and R. G. Stuart, Phys. Rev. Lett. **82** (1999) 488.
15. The MuLan Collaboration, D. B. Chitwood et al., Phys. Rev. Lett. **99** (2007) 032001.
16. The FAST Collaboration, A. Barczyk et al., Phys.Lett. **B 663** (2008) 172, arXiv:0707.3904, 2007.
17. The Particle Data Group, C. Amsler, et al., Phys. Lett. **B 667** (2008) 1.
18. C. S. Wood et al., Science **275** (1997) 1759; J. Guéna, M. Lintz and M. A. Bouchiat, physics/0412017.
19. J. S. M. Ginges and V. V. Flambaum, Phys. Rept. **397** (2004) 63.
20. P. L. Anthony, et al., Phys. Rev. Lett. **95** (2005) 081610.
21. G. P. Zeller, et al., Phys. Rev. Lett **88** (2002) 091802; G. P. Zeller, et al., Phys. Rev. Lett **90** (2003) 239902 (erratum).
22. E. A. Paschos and L. Wolfenstein, Phys. Rev. **D 7** (1973) 91.
23. K. McFarland and S. Moch, *Conventional Physics Explanation for the NuTeV* $\sin^2\theta_W$, hep-ph/0306052.
24. K. S. McFarland and C. Moch, *Conventional Physics Explanations for the NuTeV* $\sin^2\theta_W$ hep-ph/0306052; D. Mason, et al., Phys. Rev. Lett. **99** (2007) 192001.
25. A. Akhundov, *Applicability of the Formulae of Bardin and Dokuchaeva for the Radiative Corrections Analysis in the NuTeV Experiment*, arXiv:0807.2673.
26. A. D. Martin, R. G. Roberts, W. J. Stirling and R. S. Thorne, Eur. Phys. J. **C 39** (2005) 155.
27. S. R. Mishra, R. Petti and C. Rosenfeld, *A High Resolution Neutrino Experiment in a Magnetic Field for Project-X at Fermilab*, arXiv:0812.4527v1.
28. The OPAL Collaboration, G. Abbiendi, et al., Eur. Phys. J. **C 45** (2006) 1.
29. The L3 Collaboration, P. Achard, et al., Phys. Lett. **B 623** (2005) 26; The L3 Collaboration, M. Acciarri, et al., Phys. Lett. **B 476** (2000) 40.
30. S. Mele, *Measurements of the Running of the Electromagnetic Coupling at LEP*, arXiv:hep-ex/0610037v1.
31. J. F. de Troconiz and F. J. Yndurain, Phys. Rev. **D 71** (2005) 073008.
32. H. Burkhardt and B. Pietrzyk, Phys. Rev. **D 72** (2005) 057501.
33. The Muon $g-2$ Collaboration, H. N. Brown, et al., Phys. Rev. Lett. **86** (2001) 2227 ; The Muon $g-2$ Collaboration, G. W. Bennett, et al., Phys. Rev. Lett. **89** (2002) 101804; The Muon $g-2$ Collaboration, G. W. Bennett, et al., Phys. Rev. Lett. **92** (2004) 161802.
34. Reviews of the $g-2$ of the muon: J. P. Miller, E. de Rafael and B. Lee Roberts, hep-ph/0703049; F. Jegerlehner, The Anomalous Magnetic Moment of the Muon, Springer Tracts in Modern Physics, Vol 226, Springer, Berlin, Heidelberg, 2007.

35. J. Bailey et al., Nucl. Phys. **B 150** (1979) 1.
36. W. Liu et al., Phys. Rev. Lett. **82** (1999) 711;
K. Hagiwara et al., Phys. Rev. **D 66** (2002) 010001.
37. J. Schwinger, Phys. Rev. **73** (1948) 416L, and Phys. Rev. **76** (1949) 790.
38. T. Kinoshita and M. Nio, Phys. Rev. **D 73** (2006) 053007.
39. M. Davier, Nucl. Phys. Proc. Suppl. **169** (2007) 288.
40. E. De Rafael, Nucl. Phys. **B 186** (2009) 211; arXiv:0809.3085v1.
41. K. Melnikov and A. Vainshtein, Phys. Rev. **D 70** (2004) 113006;
J. Erler and G. Toledo Sanchez, Phys. Rev. Lett. **97** (2006) 161801.
42. S. J. Brodsky and J. D. Sullivan, Phys. Rev. **D 156** (1967) 1644; T. Burnett and M. J. Levine, Phys. Lett. **B 24** (1967) 467; R. Jackiw and S. Weinberg, Phys. Rev. **D 5** (1972) 2473; I. Bars and M. Yoshimura, Phys. Rev. **D 6** (1972) 374; K. Fujikawa, B.W. Lee and A. I. Sanda, Phys. Rev. **D 6** (1972) 2923; G. Altarelli, N. Cabibbo and L. Maiani, Phys. Lett. **B 40** (1972) 415; W. A. Bardeen, R. Gastmans and B. E. Laurup, Nucl. Phys. **B 46** (1972) 315; T. V. Kukhto et al., Nucl. Phys. **B 371** (1992) 567; S. Peris, M. Perrottet and E. de Rafael, Phys. Lett. **B 355** (1995) 523; A. Czarnecki, B. Krause and W. J. Marciano, Phys. Rev. **D 52** (1995) 2619; A. Czarnecki, B. Krause and W. J. Marciano, Phys. Rev. Lett. **76** (1996) 3267; G. Degrassi and G. Giudice, Phys. Rev. **D 58** (1998) 053007.
43. M. Davier, et al., *The Discrepancy Between tau and e+e- Spectral Functions Revisited and the Consequences for the Muon Magnetic Anomaly*, CERN-OPEN-2009-007, arXiv:0906.5443.
44. B. A. Shwartz, *Electroweak Physics at Low Energy*, Presentation at Lepton-Photon Conference, Hamburg, 2009; proceedings to be published.
45. D. Stöckinger, J. Phys. **G 34** (2007) R45, hep-ph/0609168; D. Stöckinger, *$(g-2)_\mu$ and Supersymmetry: Status and Prospects*, arXiv:0710.2429v1.
46. G. Montagna, et al., Nucl. Phys. **B 401** (1993) 3; G. Montagna, et al., Comp. Phys. Comm. **76** (1993) 328; G. Montagna, et al., Comp. Phys. Comm. **93** (1996) 120; G. Montagna, et al., Comp. Phys. Comm. **117** (1999) 278.
47. D. Y. Bardin, et al., Z. Phys. **C 44** (1989) 493; D. Y. Bardin et al., Comp. Phys. Comm. **59** (1990) 303; D. Y. Bardin, et al., Nucl. Phys. **B 351** (1991) 1; D. Y. Bardin, et al., Phys. Lett. **B 255** (1991) 290; D. Y. Bardin, et al., hep-ph/9412201; D. Y. Bardin, et al., Comp. Phys. Comm. **133** (2001) 229; Two Fermion Working Group Collaboration, M. Kobel et al., hep-ph/0007180; A. B. Arbuzov, et al., hep-ph/0507146.
48. O. Buchmüller, et al., *Prediction for the Lightest Higgs Boson Mass in the CMSSM using Indirect Experimental Constraints*, arXiv:0707.3447; O. Buchmüller, et al., *Likelihood Functions for Supersymmetric Observables in Frequentist Analyses of the CMSSM and NUHM1*, arXiv:0907.5568.
49. The Heavy Flavour Averaging Group, E. Barberio, et al., *Averages of b-Hadron and c-Hadron Properties at the End of 2007*, arXiv:0808.1297.
50. The WMAP Collaboration, D. N. Spergel et al., *Wilkinson Microwave Anisotropy Probe (WMAP) Three Year Results: Implications for Cosmology*, astro-ph/0603449.
51. G. Bélanger, F. Boudjema, A. Pukhov and A. Semenov, *Dark Matter Direct Detection Rate in a Generic Model with micrOMEGAs 2.2*, arXiv:0803.2360.
52. S. Heinemeyer, W. Hollik, D. Stöckinger, A. M. Weber and G. Weiglein, JHEP **0608** (2006) 052, hep-ph/0611371; S. Heinemeyer, W. Hollik and G. Weiglein, Phys. Rept. **425** (2006) 265, hep-ph/0412214; updated for 2009 data.
53. The ALEPH, DELPHI, L3, OPAL Collaborations and the LEP Higgs Working Group, Phys. Lett. **B 565** (2003) 61.
54. The TEVNPH Working Group, *Combined CDF and DØ Upper Limits on Standard Model Higgs-Boson Production with up to* 4.2 fb^{-1} *of Data*, FERMILAB-PUB-09-060-E, arXiv:0903.4001v1.
55. The ALEPH, DELPHI, L3, OPAL, SLD Collaborations, the LEP Electroweak Working Group, the SLD Electroweak and Heavy Flavour Groups, Phys. Rep. **427** (2006) 257.

56. The ALEPH, DELPHI, L3, OPAL Collaborations and the LEP Higgs Working Group, Eur. Phys. J. **C 47** (2006) 547.
57. The ATLAS Collaboration, G. Aad, et al., *Expected Performance of the ATLAS Experiment Detector, Trigger, Physics*, CERN-OPEN-2008-020, arXiv:0901.0512.

Chapter 5
The ATLAS and CMS Experiments at the LHC

The ATLAS [1] and CMS [2] experiments are the two general purpose detectors at the LHC [3]. They will measure the decay products of proton–proton collisions at up to 14 TeV. Two more detectors are installed in the LHC ring: LHCb [4] is specialised on physics with b and c quarks and ALICE [5] is dedicated to the measurement of heavy ion collisions.

The LHCb experiment will measure CP violation and rare decays of b and c hadrons in order to find indirect evidence for new physics beyond the Standard Model. Todays measurements in heavy flavour physics from B factories and the Tevatron [6, 7] are fully consistent with the CKM mechanism. Nevertheless, LHCb is probing CP violation and decays of B_d, B_s and D mesons in greater detail to possibly find new sources of CP violation or effects of new, e.g. super-symmetric, particles. At luminosities of $2 - 5 \times 10^{32}$ cm^{-2}s^{-1} which LHC will deliver to the LHCb experiment, several 10^{12} $b\bar{b}$ pairs will be produced per year. The LHCb detector is built asymmetrically around the interaction point, since b and $\bar{b}$ quarks can be measured equally well in both hemispheres. LHCb physics will cover a precise measurement of B_s oscillations especially of the mixing phase ϕ_s, the determination of $\gamma = -\arg V_{ub}$ by studying hadronic B meson decays, measurements of rare decays like $B^0 \rightarrow K^{*0}\gamma$, $B_s \rightarrow \mu^+\mu^-$, $B_s \rightarrow \phi\gamma$, and more. The LHCb detector is optimised for secondary vertex location, excellent mass resolution and particle identification to further improve the precision of CP physics with quarks.

The physics program of ALICE is dedicated to the study of QCD in extreme conditions. Collisions of heavy nuclei like Pb–Pb at $\sqrt{s} = 5.5$ TeV allow the measurement of strongly interacting matter at very high energy densities in the regime $\varepsilon \approx 1 - 100$ GeV fm^{-3}. At such energy densities a new state of matter, the quark-gluon plasma (QGP) [9], consisting of deconfined quarks and gluons is expected to occur [10]. This was first discovered at the CERN SPS [11] and further studied at RHIC [12]. Also in ALICE the QGP state will be formed for only an extremely short time in each heavy ion collision, so that the measurement of the details of the QGP will be performed by determining charged particle multiplicities, particle momentum spectra and (elliptic) flow, production (respectively suppression) of heavy quarkonia like J/ψ and Υ mesons, c- and b-quark production, as well as production of high-p_T jets and photons [13, 14]. All these serve as a probe to learn more about the formation and freeze-out mechanisms of the QGP. The LHC is

A. Straessner, *Electroweak Physics at LEP and LHC*, STMP 235, 137–165,
DOI 10.1007/978-3-642-05169-2_5, © Springer-Verlag Berlin Heidelberg 2010

planned to operate in heavy-ion mode during a few weeks per year delivering about 0.5 nb^{-1} of data to ALICE each year.

The following chapters will introduce the LHC collider and concentrate in more detail on the experimental techniques of ATLAS and CMS with focus on *pp* collision physics.

5.1 The Large Hadron Collider

The LHC [3] is installed in the LEP tunnel and accelerates bunches of protons in a ring of 26.6 km circumference from the injection energy of 450 GeV to the nominal beam energy of 7 TeV. Figure 5.1 shows the LHC underground installation. The protons are produced in a duoplasmatron device and accelerated in a linear accelerator (Linac2) to 50 MeV at a pulsed current of 180 mA. From there, the beam is injected into four Proton Synchrotron Booster (PSB) rings, ramped to 1.4 GeV, and transferred to the Proton Synchrotron (PS) where the beam energy reaches 25 GeV. In the PSB and PS the LHC bunch structure is prepared. The base structure is a sequence of 72 bunches with 25 ns spacing. Each proton bunch contains 8.28×10^{12} protons, when the LHC is operated at nominal luminosity. The rise-time of the beam ejection kicker magnet creates a gap of $(320 \approx 13 \times 25)$ ns after each 72 bunches.

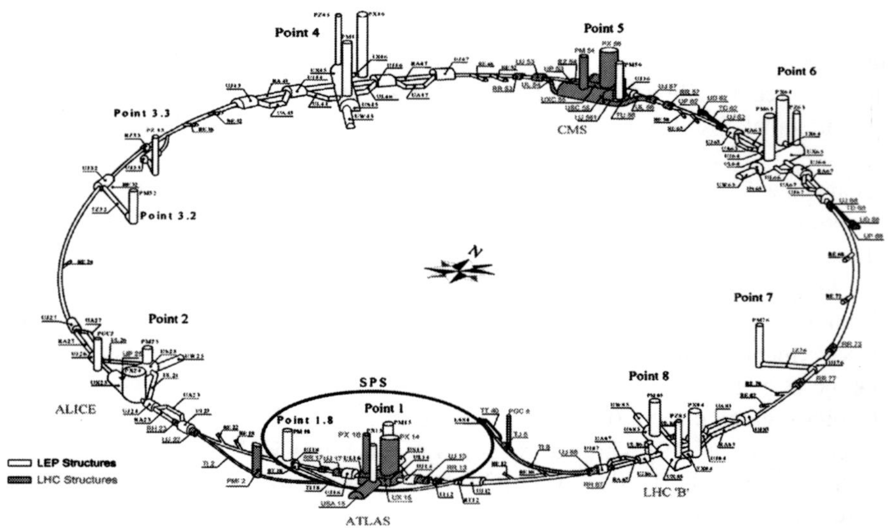

Fig. 5.1 The LHC underground installation. The ATLAS and CMS experiments are installed at opposite sites of the main ring at access points 1 and 5. ALICE and LHCb are close to the ATLAS site at point 2 and 8, respectively. The protons are injected from the SPS into the LHC via beam transfer lines. Eventually, the protons are stopped at the end of the beam lifetime into a beam dump system at point 6

The last element is the Super Proton Synchrotron (SPS) where the beam energy is increased from 25 to 450 GeV, the LHC injection energy. Additional time gaps are introduced into the final bunch structure by the rise-time of the SPS and LHC injection kickers, which is shown in Fig. 5.2. In total, 2808 bunches each containing 1.15×10^{11} protons are circulating in both directions of the LHC ring, where they are accelerated to their final energy of 7 TeV. The beam current is 0.58 A and the expected filling time is about 3 min. At the highest intensities, an energy of 362 MJ is stored in each beam.

The LHC machine is divided into eight 3 km long arc sections and eight straight sections, each 523 m long. One arc is composed of 23 identical FODO cells, with the typical focusing and defocusing magnetic multipole structure. Each cell is 107 m long with 3 dipoles and 1 quadrupole per half-cell. The superconducting dipoles provide the magnetic bending field which varies from 0.54 T at beam injection to 8.35 T when the protons reach 7 TeV. Inside the dipoles, the two proton beams circulate in two beam pipes separated by 197 mm, as displayed in Fig. 5.3. The straight sections are equipped with dispersion suppressors to match and adapt the beam optics in the straight sections, also called insertion regions (IR), to the arc. The beams are injected in IR8 and the RF structures are installed in IR4. They accelerate the protons using a 400 MHz superconducting cavity system. The frequency matches the bunch length, which varies between 1.7 and 1.1 ns at 450 GeV and 7 TeV, respectively, to capture the beam with minimal losses. Eight cavities with 5.5 MV/m field strength provide in total 16 MV accelerating voltage per beam. A photograph of one cavity is shown in Fig. 5.3.

In two other insertions the beams are cleaned: particles with large momentum offset are absorbed by collimators in IR3, while particles with large horizontal,

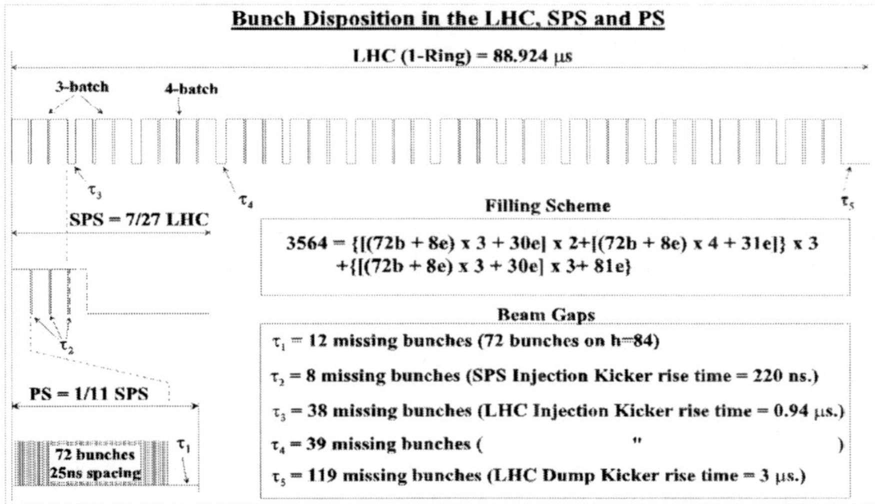

Fig. 5.2 Time structure of the LHC proton bunches. In total, 2,808 bunches are injected per proton beam [3]

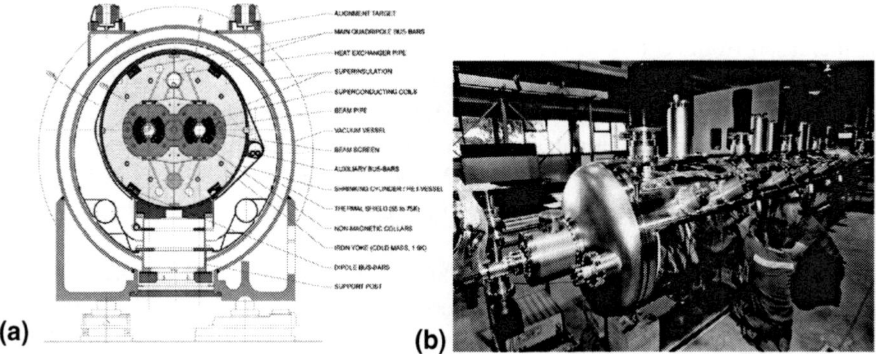

Fig. 5.3 (**a**) Cross-section through a LHC dipole magnet inside its vacuum vessel with the two beam pipes, the cold screen and the superconducting coils [3]. (**b**) One of the four superconducting cavities that are combined in each of the two RF modules which supply the 400 MHz RF power to the proton beam

vertical, or combined betatron amplitudes are filtered out in IR7. The beam abort system with the beam dump is installed in IR6.

To focus the beam at the interaction points (IP), a triplet of 31 m long quadrupole magnets is installed on each side of the IP. Beam separation and recombination is performed by two pairs of dipoles separated by 88 m. A set of four more quadrupoles provides the matching of the beam optics in the IP region to the remaining ring.

There are 1,232 main dipole magnets installed, together with several hundred quadrupoles for focusing and defocusing of the beam, completed by many thousand sextupoles, octupoles and decapoles for orbit correction. Each multipole creates betatron oscillations in the vertical and horizontal plane. In the LHC, the number of oscillations per turn, also called horizontal and vertical tune, are carefully chosen to be $Q_h = 64.31$ and $Q_v = 59.32$, respectively. This is to avoid the resonance condition

$$m Q_h + n Q_v = p \quad (m, n, p = \text{integer numbers}) \tag{5.1}$$

which leads to beam instabilities and eventually beam loss.

The machine luminosity depends on the beam parameters and can be written for a Gaussian beam distribution as

$$L = \frac{N_b^2 n_b f_{rev} \gamma_r}{4\pi \varepsilon_n \beta^*} F \;, \tag{5.2}$$

where N_b is the number of protons per bunch, n_b the number of bunches per beam, f_{rev} the revolution frequency, γ_r the relativistic gamma factor, ε_n the normalised transverse beam emittance, β^* the beta function at the IP. The factor F is the geometrical luminosity factor due to the beam crossing angle θ_c:

$$F = 1/\sqrt{1+\left(\frac{\theta_c\sigma_z}{2\sigma^*}\right)^2}\,, \tag{5.3}$$

where σ_z is the RMS bunch length and σ^* the transverse RMS beam size at the IP. The normalised transverse emittance is related to σ^* by

$$\varepsilon_n = \gamma_r \varepsilon = \gamma_r \frac{(\sigma^*)^2}{\beta^*}\,, \tag{5.4}$$

so that the luminosity may also be written in a more classical way

$$L = \frac{N_b^2 n_b f_{rev}}{4\pi(\sigma^*)^2} F, \tag{5.5}$$

inversely proportional to the transverse beam area.

Table 5.1 summarises the parameters which need to be reached to achieve the peak luminosity of 1.0×10^{34} cm^{-2}s^{-1} at 14 TeV centre-of-mass energy. With a luminosity lifetime of 15.5 h, an integrated luminosity of about 60–80 fb^{-1} per year can be expected for the final LHC performance.

There are several effects that limit the LHC luminosity. The mechanical aperture of the beam pipe is about 34.6 mm $\times$ 44 mm, and the transverse beam size is 1.2 mm, applying a 10σ safety distance of the beam profile to the wall. With a maximum value of the β function of 180 m, the normalised transverse emittance is 3.75 μm. The minimal β^* at the IP and maximum crossing angle is limited by the aperture of the quadrupole triplets.

In the interaction region, beam–beam interaction induces a so-called tune shift of:

$$\xi = \frac{N_b r_p}{4\pi\varepsilon_n} \tag{5.6}$$

Table 5.1 LHC beam parameters for 7 TeV beams at peak luminosity at the interaction points of the ATLAS and CMS experiments [3]

Number of protons per bunch	N_b	$1.15 \cdot 10^{11}$
Number of bunches	n_b	2,808
Revolution frequency	f_{rev}	11.245 kHz
Relativistic factor	γ_r	7461
Normalised transverse emittance	ε_n	3.75 μm rad
Optical beta function at IP	β^*	0.55
Full crossing angle at IP	θ_c	285 μrad
RMS bunch length	σ_z	7.55 cm
RMS transverse bunch size	σ^*	16.7 μm
Crossing angle factor	F	0.836
Luminosity per bunch	L/bunch	3.56×10^{30} cm^{-2}s^{-1}
Instantaneous luminosity	L	1.0×10^{34} cm^{-2}s^{-1}

where r_p is the classical proton radius $r_p = \alpha_{\text{QED}}/m_p$. From experience with proton-proton machines, the total linear tune shift must stay below 0.015 for stable running. For 3 IP's, this corresponds to $\xi < 0.005$, or, with $\varepsilon_n = 3.75\,\mu\text{m}$, to a maximum number of protons per bunch of $N_b = 1.5 \times 10^{11}$. This is close to the nominal value of $N_b = 1.15 \times 10^{11}$.

One important factor is the electron cloud effect. It is induced by synchrotron radiation which creates a 6.7 keV energy loss of the 7 TeV proton beam per turn. The UV photons hit electrons off the beam pipe wall. These electrons are accelerated in the field of the proton beam and can initiate secondary electron emission, which eventually builds up an electron cloud. This cloud leads to beam instabilities, growth of the emittance and increases the heat load in the cold beam screen. This screen is kept a temperature 5–20 K and shields the 1.9 K cold core of the superconducting magnets from quenches. The cooling capacity of the screen is 1.15 W/m, where the average arc heat load is already 0.66 W/m, e.g., due to synchrotron radiation and resistance of the wall. The remaining 0.5 W/m are attributed to the electron cloud heat load. The heat load is increasing also with shorter bunch spacing. For a future LHC upgrade, higher luminosity can therefore not be reached by simply increasing the number of bunches. In nominal LHC running, the electron cloud is suppressed by reducing the photon reflectivity in the arcs, and by special getter material, TiZrV, in certain sections to reduce secondary emissions. Also, the conditioning of the arc chambers by so-called "beam scrubbing", i.e. cleaning of the walls using the electron cloud effect itself, helps to reduce the disturbances due to scattered electrons during LHC operation.

At the start-up in 2009, LHC will not yet be operated at the full energy, but only at a centre-of-mass energy of initially 7 TeV, to be further increased to 10 TeV. The dipoles are expected to be commissioned to the full energy in 2011/2012. Initially, not all bunches will be filled and a 43-on-43 or a 156-on-156 bunch scenario is foreseen. The β^* in the pilot run will be in the range $3 - 11$ m, the number of protons per bunch may reach 5×10^{10}, and no beam crossing-angle is foreseen. Instantaneous luminosities in the order of $10^{31} - 10^{32}$ cm^{-2}s^{-1} can be expected in the first run period, and during 100 days of running an integrated luminosity of 100 pb^{-1} may be collected, followed by a period of another 100 days with twice the luminosity. In the following three years, a low luminosity period with $L = 10^{33}$ cm^{-2}s^{-1} is foreseen, corresponding to $1.0 - 2.5$ fb^{-1} of data in 2011/2012 assuming 150 days of physics running. Initially, there will be 936 bunches per beam with with 75 ns spacing and a 250 μrad crossing angle. Eventually, the nominal number of bunches is increased to 2,808 with 25 ns spacing and nominal 285 μrad crossing-angle. The number of protons per bunch will stay in the order of 5×10^{10}.

After the full implementation of the collimators and the final completion beam dump system in 2013/2014, the beam intensities can be pushed to 9×10^{10} protons per bunch and the β^* will be squeezed to 0.55 m. With these parameters, the design luminosity of $L = 10^{34}$ cm^{-2}s^{-1} is expected to be reached and the LHC will be operated in this mode until 2017/2018 with up to $80 - 100$ fb^{-1} of data per year.

From the above discussion one can see that possible ways to further upgrade the LHC machine to higher luminosity are, for example, a modified triplet magnet with larger aperture and smaller $\beta^* = 0.25\,\mu$m and/or longer proton bunches with larger bunch spacing. This luminosity upgrade, with a possible increase of the beam energy, is foreseen for a shutdown period in 2018/2019. In the most complete scenario, this requires a replacement of the complete injection chain to attain higher proton energies and better beam brilliance N_b/ε^* at the interaction points.

The actual proton–proton interaction rate seen by the CMS and ATLAS detectors in the interaction regions is proportional to the luminosity and given by

$$\frac{dN}{dt} = L\sigma_{pp} , \tag{5.7}$$

where the main contribution to the proton–proton cross-section, σ_{pp}, is from inelastic scattering, σ_{inel}. It amounts to about 80 mb, or 80×10^{-28} cm^2. This means at the peak luminosity of 10^{34} cm^{-2}s^{-1}, the number of events per bunch crossing, N_c, is about

$$N_c = \frac{L}{n_b}\frac{\sigma_{inel}}{f_{rev}} = 25.6 . \tag{5.8}$$

On top of each hard pp scattering event, for example the production of a Z boson in a Drell-Yann process, 25 inelastic pp events are overlaid. This phenomenon of in-time pile-up, the high collision frequency and very good resolution for optimal signal identification and background rejection drives the performance requirements of the ATLAS and CMS detectors.

5.2 The ATLAS and CMS Experiments

The layout of the ATLAS and CMS experiments follows the well-known concepts of particle collider detectors with tracking systems for charged particles close to the interaction point, electromagnetic and hadronic calorimeters at larger distances, completed by muon detectors in the outermost layer. A schematic view of ATLAS and CMS is shown in Figs. 5.4 and 5.5. Both detectors provide maximal hermeticity in their angular coverage. The general design is furthermore motivated by the challenging physics tasks of experiments at the LHC. The proton–proton beam crossing rate is at 40 MHz, each crossing with up to 25 inelastic pp collisions. This results in a high density of charged particle tracks, calorimetric energy deposits and signals in the muon detectors. For example, the flux of charged particles at a radius $r = 10$ cm around the beam axis is about 10 MHz/cm^2. Apart from providing a good energy and momentum resolution for leptons, jets and global energy flow, the detection systems must in addition be of high granularity and must sustain high radiation levels. They should also have a fast signal response time and be equipped with a fast electronic readout to filter out the interesting physics processes using a multi-layer trigger system.

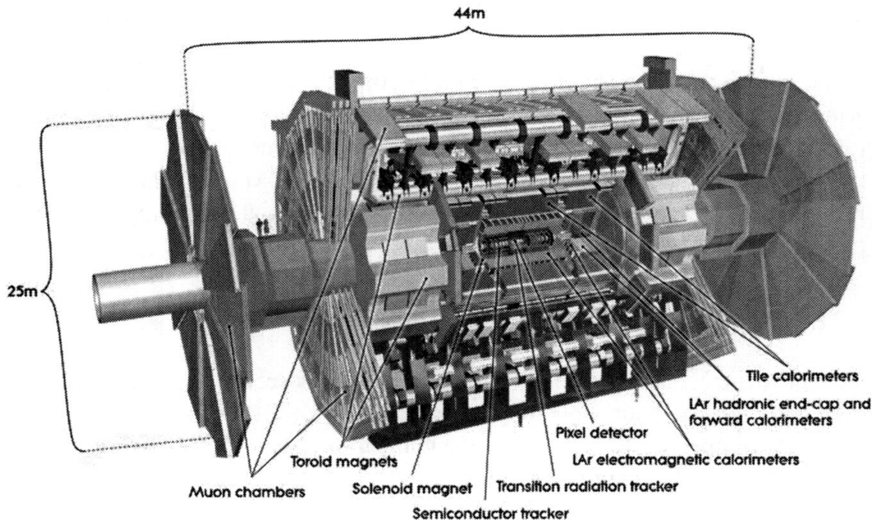

Fig. 5.4 Cut-away view of the ATLAS detector. The dimensions of the detector are 25 m in height and 44 m in length

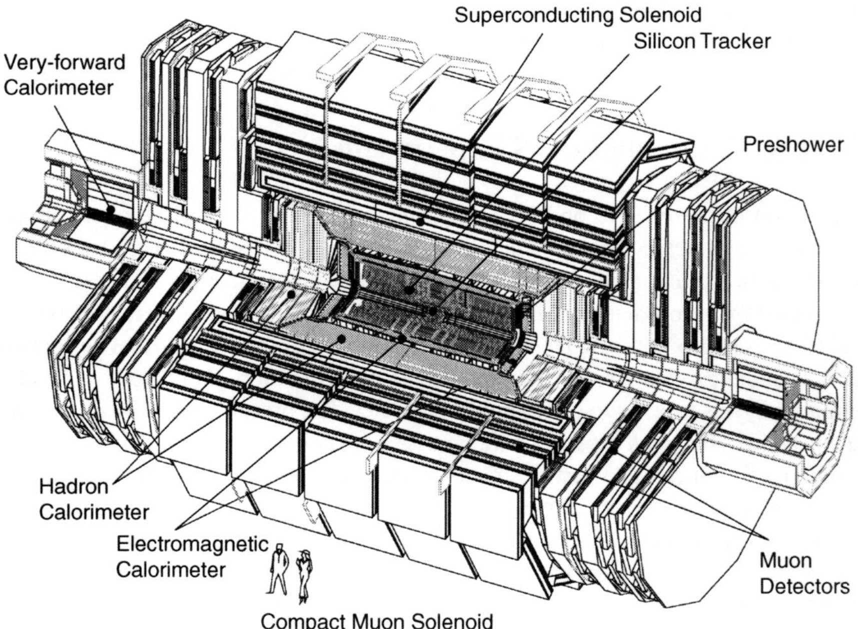

Fig. 5.5 Schematic view of the CMS detector showing the compact design of the different particle detection systems

5.2.1 The ATLAS Detector and Performance

The ATLAS detector [1, 8] is located at IP 1 of the LHC ring. Closest to the collision point is the inner detector for tracking of charged particles, which is installed in a cylindrical superconducting solenoid of 5.5 m length and 1.15 m radius. The particles are bent in a 2 T magnetic field and detected by three separate tracking devices.

The silicon pixel detector is composed of three barrel layers at radii between 50.5 and 122.5 mm and three endcap disks at 495–650 mm distance to the nominal interaction vertex. The size of the 47,232 pixels is $50 \times 400\,\mu\text{m}^2$ and the intrinsic spacial resolution is $10\,\mu\text{m}$ and $115\,\mu\text{m}$ in $r - \phi$ and z/r direction, respectively. The subsequent tracking is done by the silicon tracker (SCT) which has 4 cylindrical silicon microstrip layers in the barrel section and 2×9 microstrip disks in the endcaps. The outermost silicon layer is at 514 mm radius and the most distant disk at $|z| = 2,727$ mm. With a strip width of $80\,\mu\text{m}$ in the barrel and 57–94 μm in the endcap, a spacial resolution of 17 and $580\,\mu\text{m}$ in $r - \phi$ and z/r direction can be achieved. The combined angular coverage of the pixel and SCT reaches to $|\eta| < 2.5$. Further details of the geometrical layout are shown in Fig. 5.6

The inner detector is completed by a transition radiation tracer (TRT) which uses straw tubes filled with a $Xe/CO_2/O_2$ gas mixture as active medium. The tubes have a 4 mm diameter and a length of 144 cm in the barrel and 37 cm in the endcap. They provide additional 20–35 space points along the particle tracks in the fiducial volume $|\eta| < 2.0$. The combined momentum resolution for single charged tracks in the inner detector, using pixel, SCT and TRT, is

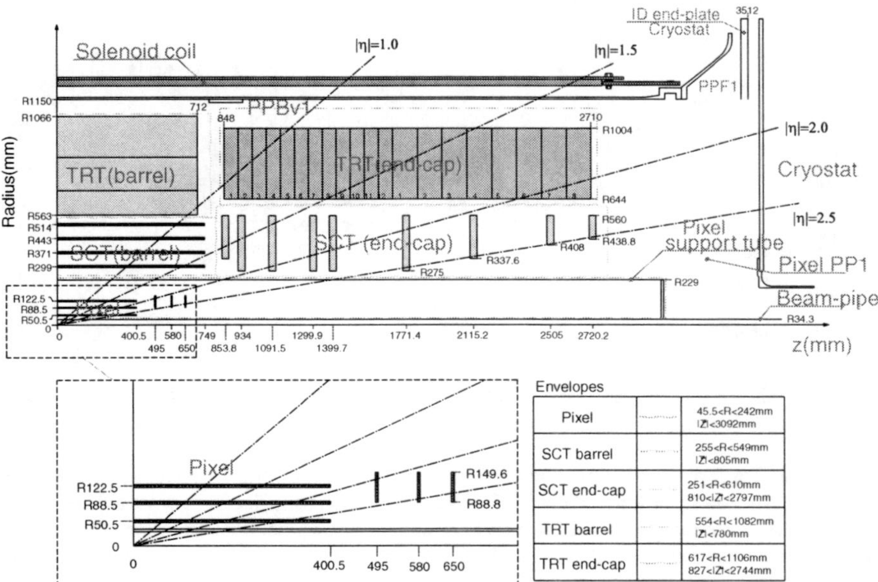

Fig. 5.6 Geometrical layout in the $r - z$ plane of one quarter of the ATLAS inner detector with the pixel, SCT and TRT sub-detectors [1]

$$\frac{\sigma_{p_T}}{p_T} = \sqrt{\left(5 \cdot 10^{-4} p_T\right)^2 + 0.01^2} \tag{5.9}$$

in the fiducial volume $|\eta| < 2.5$.

The TRT is furthermore used to separate electrons and pions. Photon radiation is produced at the transition from the straw tube plastic and the gas, an effect that depends on the relativistic γ of the particle, which, for the same energy, is about 280 times larger for electrons than for pions. The photons created by traversing electrons are in the X-ray energy range and produce additional high threshold signals in the straw tube detectors. The high threshold probability is used as a discriminant against pions. For particles of 25 GeV, a $\pi^{\pm}$ rejection factor between 10 and 100 at an electron efficiency of 90% is obtained.

A serious challenge for the measurement of electromagnetic particles is the number of interaction lengths in the silicon tracker in front of the calorimeter. About 20–60% of the high-energy photons undergo conversion into an e^+e^- pair, while electrons loose between 30 and 60% of their energy due to bremsstrahlung. This is

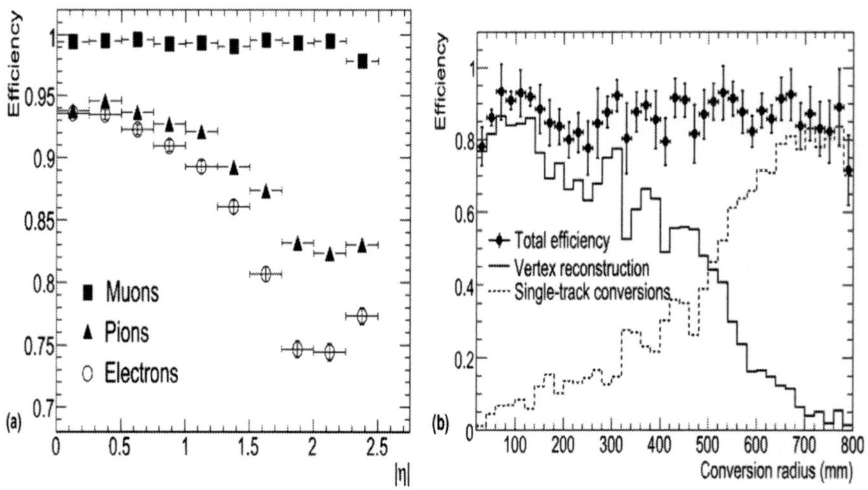

Fig. 5.7 (**a**) Track reconstruction efficiency in the inner detector for 5 GeV electrons, pions and muons [1]. The lower tracking efficiency for electrons and pions is due to bremsstrahlung effects when traversing the material of the inner detector and the beam pipe. It is increasing with η. Minimum ionising muons are less affected. (**b**) Efficiency to reconstruct photon conversions of photons with $p_T = 20$ GeV in the central $|\eta| < 2.1$ detector region, as a function of the conversion radius [1]. The overall efficiency is a combination of double and single track conversions

compensated by taking the two effects into account during reconstruction. Figure 5.7 gives examples for the track reconstruction efficiencies for electrons and pions which are affected by bremsstrahlung as compared to muons which are minimum ionising. Also shown is the high 80% efficiency to reconstruct photon conversions. The precision of the position of the conversion vertex is about 5–7 mm in radial direction.

Vertexing precision is as well important in the identification of primary and secondary vertices. The primary vertex is reconstructed in $H \to \gamma\gamma$ events with 96% efficiency and the correct vertex is selected with 79% probability, both evaluated for a luminosity of $L = 10^{33}$ cm^{-2}s^{-1}. Vertex reconstruction efficiency and correct vertex assignment are both close to 100% for $t\bar{t}$ events where additional tracks are produced in the hard scattering process. The primary vertex resolution is about 20–35 μm in the plane transverse to the beam and 40–70 μm in longitudinal direction. The secondary vertex reconstruction efficiency depends even stronger on the event topology. For $t\bar{t}$ events it is typically higher than 60%.

The energy measurement of electrons and photons is performed in the electromagnetic calorimeter that consists of a Liquid Argon barrel and endcap calorimeter in the pseudo-rapidity ranges $|\eta| < 1.475$ and $1.375 < |\eta| < 3.2$, respectively. The calorimeter is a LAr-lead sampling calorimeter with accordion-shaped kapton electrodes and lead absorbers. The accordion structure is chosen in order to achieve a homogeneous energy response in ϕ without detection gaps. The ATLAS calorimetric devices, including the LAr calorimeters, are shown in Fig. 5.8.

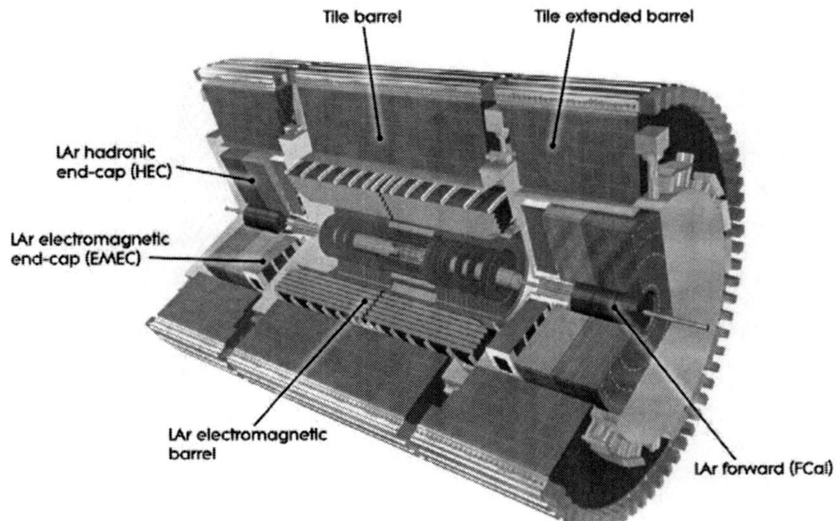

Fig. 5.8 The ATLAS Liquid Argon calorimeters consist of the electromagnetic barrel and endcap calorimeters, a hadronic endcap and a forward calorimeter. They are installed in metal cryostats onto which the front-end electronics is mounted. The ATLAS calorimetry is completed by a hadronic Tile calorimeter in the barrel and forward region

The two halves of the LAr barrel are installed in the same cryostat as the solenoid magnet and are separated by a small 4 mm gap. The barrel electrodes are segmented in radial direction in three compartments each having different signal cell sizes. The innermost layer is composed of strips covering $\Delta\eta \times \Delta\phi$ segments of $0.025/8 \times 0.1$ in the centre, $|\eta| < 1.40$, becoming more course, 0.025×0.025, in the $1.40 < |\eta| < 1.475$ region. The second layer covers 0.025×0.025 and 0.075×0.025 segments in the two angular regions, while the third layer has a granularity of 0.050×0.025. The strip layer is meant to sample the beginning of the electromagnetic shower with high resolution to be able to resolve adjacent showers, for example from $\pi^0 \to \gamma\gamma$ decays, converted photons or bremsstrahlung photons close to electron clusters. The middle layer is generally containing the peak of the electromagnetic energy deposition along a photon or electron shower, while the back layer is measuring the shower tail. An example of the LAr cells in the central barrel section is given in Fig. 5.9.

The $\Delta\eta \times \Delta\phi$ segmentation of the endcap calorimeter is similar to the barrel, with cell sizes between $0.025/8 \times 0.1$ to 0.1×0.1 in the first layer, between 0.025×0.025

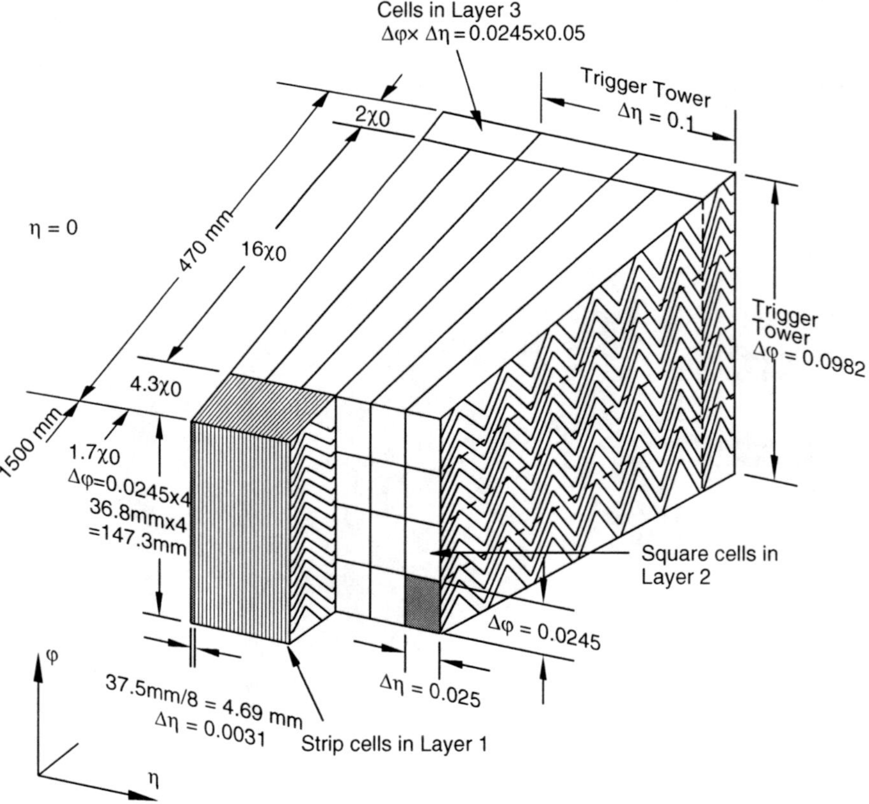

Fig. 5.9 Segmentation in $\Delta\eta \times \Delta\phi$ and radial depth of the LAr calorimeter cells in the barrel. Also indicated is the size of the calorimeter trigger towers which combine larger angular areas [1]

to 0.1×0.1 in the second layer, both varying with η, and eventually 0.050×0.050 in the back layer. The LAr calorimeter has in total about 180 thousand cells that are read out.

The barrel calorimeter has a depth of at least 22 radiation lengths, X_0, increasing to maximal 36 X_0, while the endcap provides between 24 X_0 and 38 X_0 of absorbing material. Since showering of electromagnetic particles usually starts already in the upstream material of the inner detector, a presampler is installed in front of the barrel calorimeter and in parts of the endcap. The energy deposition of photons and electrons in the 11 mm LAr gap in the barrel and the two 2 mm gaps in the endcap are proportional to their upstream energy loss. In this way the complete longitudinal shower development can be reconstructed.

Using the energy measurements in the presampler, E_{PS}, in the strip, middle and back compartments, E_{strip}, E_{middle}, E_{back}, the electromagnetic cluster energy can be reconstructed as a weighted sum:

$$E = s(\eta)\left(c(\eta) + w_0(\eta)E_{PS} + E_{strips} + E_{middle} + w_3(\eta)E_{back}\right) . \qquad (5.10)$$

The overall scale factor s, the offset, c, and the weights, w_0 and w_3, are determined in the calibration procedure and are η dependent. The weight w_0 applied to the presampler measurement is in the order of 20 and 60 for the barrel and endcap, respectively. This compensates the energy loss in front, while the weight w_3 takes the leakage of the shower at the back of the calorimeter into account. The obtained energy resolution for electrons and photons is shown in Fig. 5.10. It follows a parameterisation

$$\left(\frac{\sigma}{E}\right)^2 = \left(\frac{S}{\sqrt{E}}\right)^2 + C^2 , \qquad (5.11)$$

with a stochastic term S, and a constant term C. In the most central η region the values are $S = 10.0\%$, and $C = 0.7\%$. The latter is mainly determined by the long-range uniformity of the calorimeter and by the calibration at cell level. It also assumes perfect knowledge of the material distribution in front of the calorimeter.

The calorimetry of ATLAS is completed by a hadronic tile calorimeter in the barrel and extended barrel regions at $|\eta| < 1.0$ and $0.8 < |\eta| < 1.7$, respectively. The calorimeter is built of steel as absorber and scintillating tiles are the active material. The inner and outer radius are 2.28 and 4.25 m. At $\eta = 0$, the detector thickness corresponds to 7.4 hadronic interaction lengths, λ. The tile calorimeter is radially segmented in four layers and the cell sizes in η and ϕ direction are 0.1×0.1. The combined LAr and tile calorimeter performance for pions was determined in test beam measurements. For a calorimeter slice that corresponds to the region at $\eta = 0.25$ a pion energy resolution with a stochastic term of 52% and a constant term of 3% was found, following equation 5.11.

In the forward region, a LAr hadronic endcap calorimeter (HEC) at $1.5 < |\eta| < 3.2$ is built of 25 and 50 mm thick copper plates interleaved with LAr gaps. The granularity of the calorimeter is 0.1×0.1 for $|\eta| < 2.5$ and twice as large in

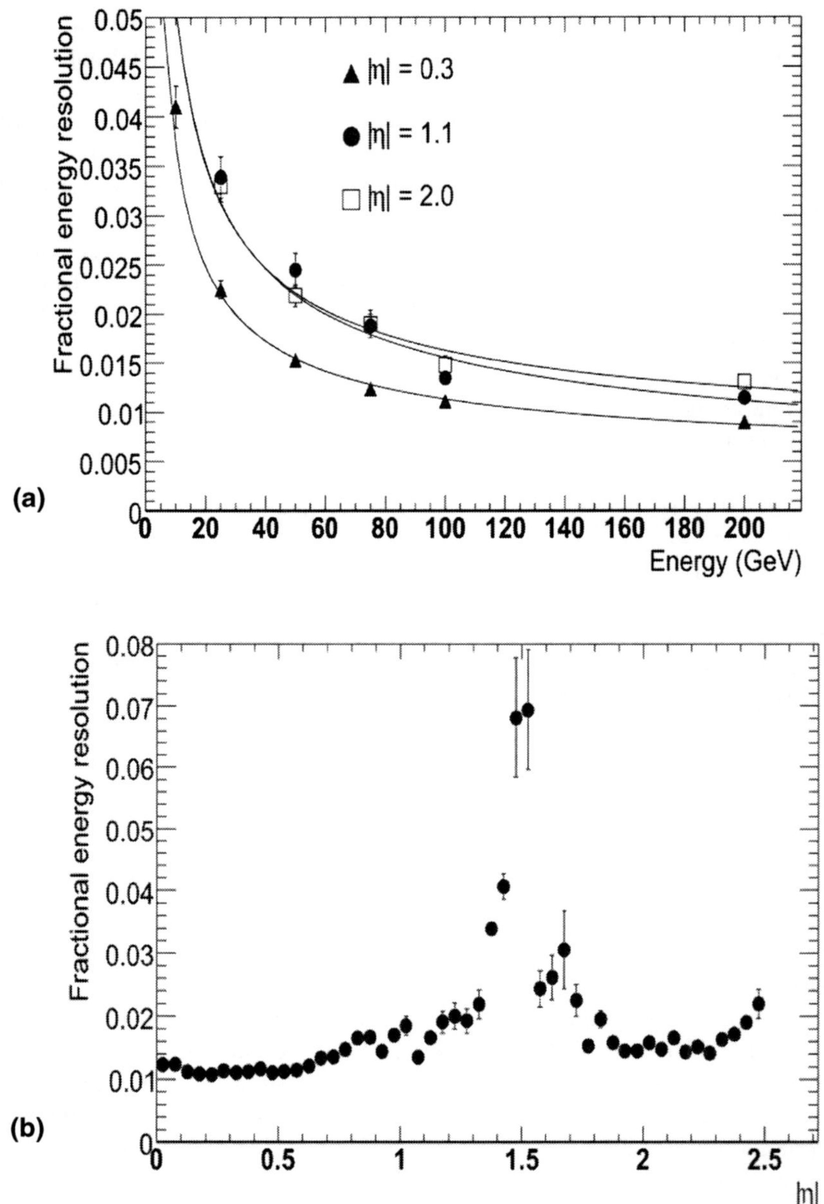

Fig. 5.10 (**a**) Expected energy resolution for electrons in different η ranges as a function of energy [1]. (**b**) Relative energy resolution for 100 GeV electrons as a function of η. The resolution is rather uniform except in the transition region between barrel and endcap [1]

the remaining η range. The LAr forward calorimeter (FCAL) further extends the angular coverage to $3.1 < |\eta| < 4.9$. It consists of three modules with in total 10 λ depth. The one closest to the IP is intended for electromagnetic measurements and made of copper, while the other two are made of tungsten and measure hadronic particles. The FCAL energy resolution for electrons follows the functional form of Eq. 5.11 with $S = (28.5 \pm 1.0)\%$ and $C = (3.5 \pm 0.1)\%$, which is obtained in testbeam measurements. The pion energy resolution is about a factor 2–3 worse, depending on the sophistication of the reconstruction algorithm. The stochastic term of the global jet energy resolution is therefore expected to be 50% for $|\eta| < 3.2$ and 100% for $3.1 < |\eta| < 4.9$, while the constant energy resolution term is 3 and 10%, respectively.

Jets are reconstructed from calorimeter towers with $\Delta\eta \times \Delta\phi = 0.1 \times 0.1$ cell size or from so-call topological clusters. The latter are reconstructed from combined high resolution calorimeter cells and noise is already subtracted. The basic jet clustering algorithms are the seeded iterative cone [15] and the k_T algorithm [16]. The former combines the input objects in angular cones of a fixed sizes $\Delta R = \sqrt{\Delta\eta^2 + \Delta\phi^2}$, with $\Delta R = 0.4$ and 0.7 for narrow and wide jets with a seed energy of $E_T = 1$ GeV. It is the most widely used algorithm in ATLAS, but it is neither infrared nor collinear safe and may lead to inconsistencies with fixed order QCD calculations. More performing algorithms like SISCone [17] or anti-k_T [18], which theoretically preferred since they are infrared and collinear safe, are under study. Details of the standard cone jet resolution are shown in Fig. 5.11. The stochastic term is in the order of 60% and the constant term around 3%. The noise contribution is 0.5 GeV in the barrel and increases to 1.5 GeV in the endcap. The reconstruction efficiency approaches 100% at jet p_T values of more than 40 GeV. The jet direction is determined to better than $\Delta R = 0.2$ for jets with $p_T > 100$ GeV.

Also important is the reconstruction of the missing transverse energy, E_T^{miss}, which provides signatures for particles that escape detection, like neutrinos or non-interacting super-symmetric particles. It is calculated as the E_T sum of the calorimetric towers and also reconstructed muons are taken into account. The corresponding resolution is shown in Fig. 5.11 as a function of $\sum E_T$ using different simulated physics processes. It can be parameterised at high $\sum E_T$ by $\sigma_{E_T^{miss}} \approx 0.57 \cdot \sqrt{\sum E_T}$. The direction of the missing E_T vector in the $x - y$ plane is determined with a precision better than $\Delta\phi = 0.8$ at low E_T^{miss} reducing to below 0.1 at $E_T^{miss} > 150$ GeV.

The outermost detector layer of ATLAS is composed of four different muon detection systems: monitored drift tubes (MDT) in barrel and endcap for the precise measurement of the muon momenta, thin-gap chambers (TGC) in the endcap for triggering, cathode-strip chambers (CSC) for the innermost endcap region, and eventually resistive plate chambers (RPC) for triggering and momentum determination in the barrel. The general geometrical arrangement of the muon chambers can be seen in Fig. 5.4 and a side view in the $r - z$ plane is shown in Fig. 5.12. The angular coverage of the MDT is $|\eta| < 2.7$ and there are more than 1,000 chambers with 340 thousand channels installed. The innermost MDT layer only reaches to $|\eta| < 2.0$. CSC chambers cover this endcap region $2.0 < |\eta| < 2.7$ with high neutron background and high particle rate. The more than 500 RPC trigger chambers

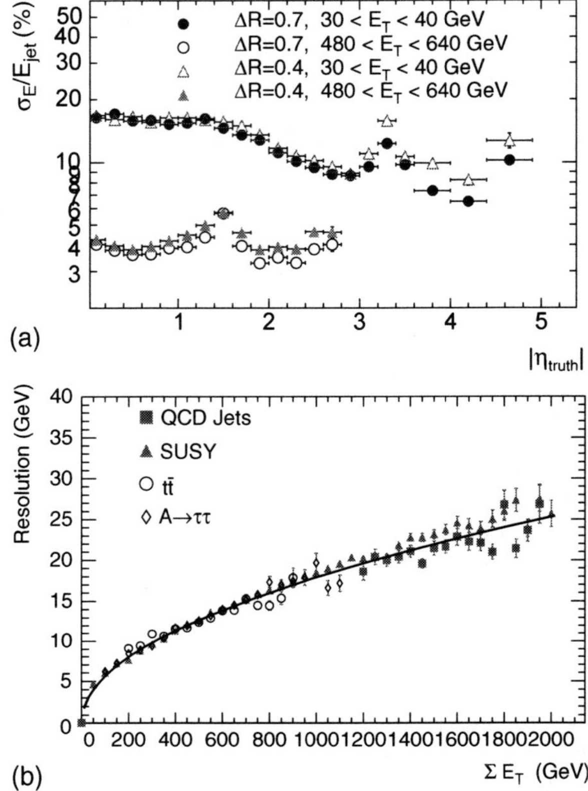

Fig. 5.11 (**a**) Expected jet energy resolution for seeded iterative cone jets with different cone size and energy range [1]. (**b**) Resolution of the reconstructed transverse missing energy as a function of the sum of the transverse energy in the event [1]. Different event samples are used to evaluate the resolution at various transverse energy scales

are installed at the middle and outer MDL layers in the fiducial region $|\eta| < 1.05$. Triggering is extended to $1.05 < |\eta| < 2.7$ in the forward region by 3,588 TGC chambers.

The magnetic bending field is provided by three air-coil superconducting toroid magnets. Eight barrel toroid coils create a field of 0.15–2.5 T, while the endcap field is between 0.2 and 3.5 T, depending on the azimuthal and polar angle. The endcap toroid also has an eight-fold symmetry and is rotated by $\pi/8$ with respect to the barrel coils. The analysing power is better quantified in terms of the field integral along the possible muon trajectory, which is displayed in Fig. 5.13. It shows that the field integral is mostly between 2 and 7 Tm, except in the barrel-endcap transition region where the momentum reconstruction power is much reduced.

The resolution for high momentum muons is shown in Fig. 5.13. At very high momentum of $p_T = 1$ TeV the relative resolution is expected to be on the order of 10%.

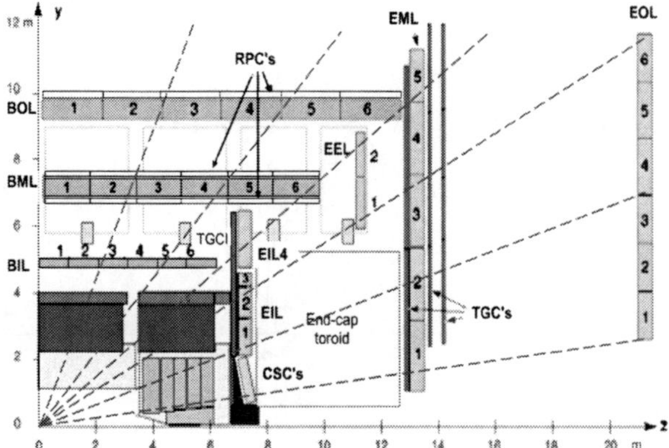

Fig. 5.12 Schematic view in the $r-z$ plane of one quarter of the ATLAS muon detectors [1], with MDT chambers in barrel and endcap inner, middle and outer layer (BIL, BML, BOL, EIL, EML, EOL), the CSC and TGC detectors in the endcap region, as well as the RPC trigger layers attached to the barrel middle and outer MDTs

For many of the projected physics analyses, in particular for measurements of cross-sections, the knowledge of the absolute, and eventually integrated LHC beam luminosity is needed. The main luminosity detector of ATLAS is the Cerenkov integrating detector LUCID [19, 1] which is installed at 17 m distance to the ATLAS interaction point. It will measure the number of elastic pp interactions at every bunch crossing, which is proportional to the LHC luminosity. An absolute calibration of the luminosity is provided by the ALFA [19, 20] system which consists of scintillating-fibre trackers located inside Roman Pots at a distance of 240 m from the interaction point.

The measurements with the ALFA detector require dedicated runs with special beam optics and low luminosity, $L = 10^{27}\ \mathrm{cm}^{-2}\mathrm{s}^{-1}$. The LUCID detector will be operated in parallel so that the two will be inter-calibrated. LUCID is then used to extrapolate this measurement up to the design luminosity with a final expected accuracy of 2–3%.

5.2.2 The ATLAS Trigger and Data Acquisition System

Important for physics measurements is the trigger system. The pp bunch crossing rate is at 40 MHz and the inelastic cross-section about 80 mb. That means that at high luminosities of $10^{34}\ \mathrm{cm}^{-2}\mathrm{s}^{-1}$ there are about 25 interactions per bunch crossing. The interesting physics processes have however much lower cross-sections in the picobarn to below femtobarn range. In general, high transverse momentum electrons, photons, muons, taus, and jets, as well as large missing transverse momentum are typical signatures of the physically interesting hard scattering processes.

The ATLAS trigger system is divided into three layers, where the first one (L1) is implemented in custom hardware and the second and third level triggers, L2 and

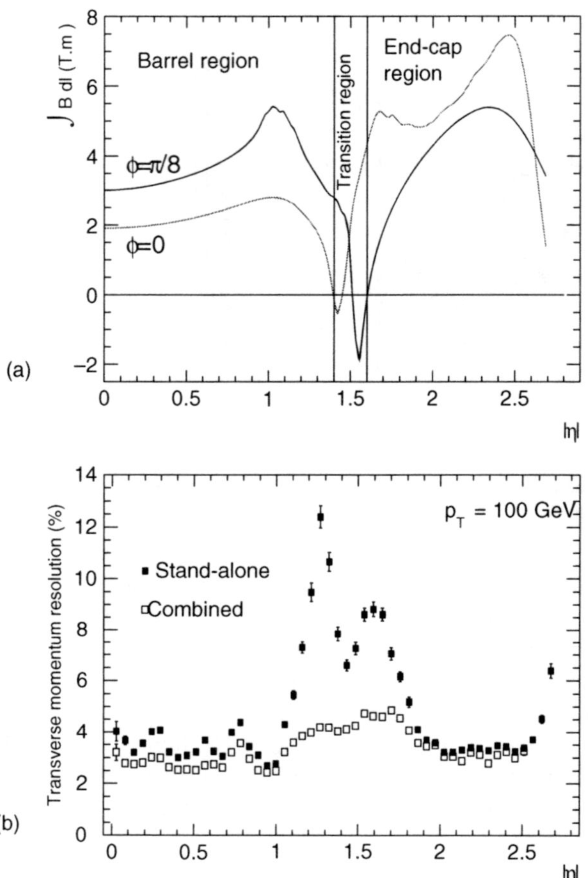

Fig. 5.13 (**a**) Integral of the magnetic field seen by muons along their trajectory [1]. To visualise the effect of the eight-fold symmetry of the toroid system the integral is evaluated at $\phi = 0$ and $\phi = \pi/8$. (**b**) Momentum resolution for $p_T = 100$ GeV muons in different η regions [1]. The stand-alone resolution is mostly better than 4%, exccept in the barrel-endcap transition region and in the very forward η range. If the muon chamber reconstruction is combined with inner detector tracks the resolution is generally improved, in particular in the transition region

Event Filter (EF), are based on software algorithms running on large PC farms. The software triggers are also called high-level triggers (HLT).

The first level trigger has two main inputs: the calorimeter trigger and the muon trigger. The corresponding trigger signals are read out with an electronics chain parallel to the standard readout. The data from the detector front-end is already concentrated so that a fast trigger decision with a maximum trigger latency of 2.5 μs is provided. After this time interval the readout buffers on the detectors must have received a L1 signal to transfer the detector data to the Data Acquisition system (DAQ). The front-end buffer sizes are chosen according to this timing structure.

The calorimeter trigger receives information from about 7,000 analogue trigger towers with mostly $\Delta\eta \times \Delta\phi = 0.1 \times 0.1$ granularity. After digitisation, a Cluster Processor (CP) and a Jet/Energy-sum Processor (JEP) identify physics signatures. The CP looks for electron, photon and τ candidates above programmable p_T thresholds with possible isolation from other detector activities, while the JEP forms jets, and calculates the quantities $\sum E_T$ and E_T^{miss}, which also have to fulfil energy and multiplicity thresholds.

The muon trigger system in the barrel region is identifying low p_T muons from hits in both innermost RPC layers that are mounted inside and outside of the middle MDT stations. The p_T measurement is performed using pre-defined tracking roads whose width selects different p_T values: more narrow roads correspond to higher p_T thresholds. For high p_T tracks in the innermost RPC layers, a third measurement in the outer RPC layers is used for refining the threshold using a similar algorithm based on tracking roads. In the endcaps, the TGC chambers provide the trigger input. They stand higher rates than RPCs and operate with 99% efficiency up to rates of $20\,\text{kHz/cm}^2$. Coincidence hits in the outer TGC chambers are treated independently in r and ϕ direction. They are finally merged and combined with the innermost TGC chamber hits. Six geometrical windows for the hits along the muon track correspond to pre-defined p_T thresholds, similarly to the barrel algorithm.

The L1 trigger decisions are taken by the central trigger processor at a rate of 75 kHz. On a L1-accept signal the detector data are transfered via optical fibres from the front-end through the Read-Out Drivers (ROD) to the DAQ system and the subsequent L2 algorithms are started. Each L1 object defines a so-called region of interest (ROI) which is an angular cone around the η and ϕ direction of the identified particle or E_T^{miss} candidate. The L2 algorithms are seeded from the ROI and only accesses data within this region, which is about 1–2% of the full event information. The L2 algorithms search for more refined physics signatures, so-called trigger chains. These chains correspond to step-wise hypothesis tests to verify if the detected signature corresponds to the programmable L2 selection criteria. In each step, the hypotheses become more restrictive. Trigger signatures are rejected as early as possible, whenever a hypothesis step is not fulfilled any more. Typical trigger signatures are

- electrons reconstructed as electromagnetic calorimetric clusters matched to inner detector tracks,
- photons identified as electromagnetic calorimeter clusters without matched track,
- tau leptons with hadronic decay signatures in the calorimeters and matched tracks,
- cone jets formed from calorimeter towers, including b-tag information,
- muons reconstructed from combined tracks in the muon chamber and the inner detectors,
- transverse calorimetric energy sums, and
- missing transverse energy.

For all trigger objects, energy and p_T threshold can be freely defined as well as isolation from nearby hadronic activity, which is important for background rejection especially at high luminosities. The L2 trigger decision must be taken within 40 ms.

The last trigger level, the event filter (EF), has access to the complete event data. These data are provided at a maximum frequency of 3.5 kHz. The EF is able to perform event selections very similar to offline data analysis so that more complicated signatures can be implemented. The EF algorithms are tuned such that the final event output rate is at most 100 Hz. The events are grouped in different categories or data streams. Typical event rates at luminosities of 10^{33} cm^{-2}s^{-1} in the different streams are

- electron stream: 31 ± 8 Hz
- muon stream: 34 ± 9 Hz
- jet stream: 38 ± 6 Hz
- E_T^{miss} and τ stream: 32 ± 8 Hz
- B-physics stream: 10 ± 6 Hz

In this way, the input to the offline data analyses can be reduced to the selected trigger streams so that total data processing time can be optimised for the various physics analyses.

5.2.3 The CMS Detector and Performance

The CMS detector [2] has a length of 21.6 m and a diameter of 14.6 m and is installed at IP 5 of the LHC ring. The central tracker, as well as the electromagnetic and the hadronic calorimeters are inside a solenoidal magnetic field of 4 T in which the charged particle tracks are bent. The solenoid is made of a superconducting coil of 5 m radius and 13 m length. The iron return yoke on the outside of the solenoid is equipped with muon chambers. CMS has thus a much more compact design than the ATLAS detector.

The CMS inner tracking system is composed of a silicon pixel and microstrip detectors. The central barrel region is covered by 3 pixel layers and 9 microstrip layers. The barrel region is completed with 2 pixel disks and 3 microstrip disks oriented perpendicular to the beam axes. The tracker endcap consists of 9 microstrip disk layers, which are installed parallel to the barrel disks. In the central barrel region, the innermost strip layers and two strip layers of the outer barrel are made of two-layer modules with a stereo angle of 100 μrad to provide measurement in the $r - \phi$ and $r - z$ planes. There are in total 66 million pixels and 9.6 million silicon strips, which cover an angular region up to $|\eta| < 2.4$. The occupancy in the $100 \times 150\,\mu\text{m}^2$ large pixels is 10^{-4} per beam crossing, while in the inner $10\,\text{cm} \times 80\,\mu\text{m}^2$ large strips the occupancy is 2–3%. The single-point resolution in the inner stereo layers is 23–34 μm in $r - \phi$ and 230 μm in z direction. In the outer layers it is about a factor 2 larger. The expected track resolution obtained for single muon events is shown in Fig. 5.14. For low momentum tracks below 10 GeV, the p_T resolution will be better than 0.7% in the central region and decreases to 2% for $\eta = 2.4$. Higher energy tracks of 100 GeV can be measured with 1.5–6% resolution. Another performance parameter, important e.g., for the lifetime tag of B meson decays, is the resolution on the impact parameter distance to the vertex. It is better than 200 μm for low energy tracks and reaches 10 μm for more straight tracks above 100 GeV, also shown in shown in Fig. 5.14. Some performance parameters of the ATLAS and CMS tracking systems [21] are compared in Table 5.2. The CMS tracker achieves a better momentum resolution as well as better impact parameter (IP) resolutions at high p_T due to the higher magnetic field and the smaller pixel sizes, respectively.

The CMS electromagnetic calorimeter (ECAL) is composed of a barrel and two endcaps with 61,200 and 2 × 7,324 lead tungstate ($PbWO_4$) scintillating crystals, respectively. The crystals have a short radiation length, $X_0 = 8.9$ mm, fast light response and are radiation hard. The barrel section has an inner radius of 129 cm and covers an η range up to 1.479, while the endcaps are at 3.14 m distance to the IP and cover $1.479 < |\eta| < 3.0$. The crystal length in barrel and endcap correspond to about 25 X_0, so that electromagnetic showers are contained in the calorimeter up to high energies. In the endcap region a preshower device made of 2 lead absorber disks and 2 planes of silicon strips with 1.9 mm pitch provides additional rejection power against π_0 background. The electron resolution of an ECAL module is shown in Fig. 5.15. It is parameterised as

$$\left(\frac{\sigma}{E}\right)^2 = \left(\frac{S}{\sqrt{E}}\right)^2 + \left(\frac{N}{E}\right)^2 + C^2 , \qquad (5.12)$$

with a stochastic term S, noise N, and a constant term C. The expected performance parameters are also given in Fig. 5.15. The overall resolution for a 100 GeV electron is about 0.5%. Table 5.3 compares the CMS ECAL to the ATLAS LAr calorime-

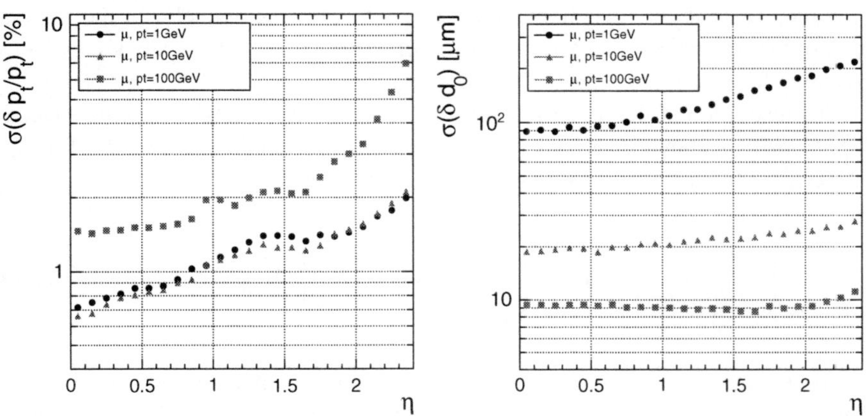

Fig. 5.14 Expected momentum and transverse impact parameter resolution of single muon tracks of three different energies, simulated for the CMS tracking system in different η regions [23]

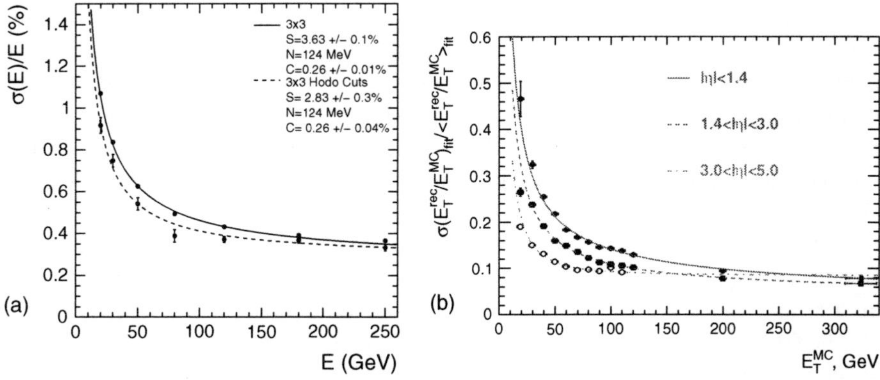

Fig. 5.15 (**a**) Electron energy resolution of a CMS ECAL module [23] measured in a beam test. "Hodo cuts" refers to a measurement using a beam hodoscope to restrict the incoming beam to the centre of the 3x3 electromagnetic cluster, which results in a better resolution since effects of energy loss at the crystal edges are avoided. (**b**) Jet transverse energy resolution for the three η regions of the hadron calorimeter, barrel, endcap and forward [23]. The jets were reconstructed using an iterative cone algorithm with $R = 0.5$. The reconstructed transverse energy, E_T^{rec}, is compared to the MC generated energy, E_T^{MC}

ter [21], where the latter shows a slightly worse performance in terms of energy resolution.

The hadronic calorimeter (HCAL) must operate in a magnetic field and is optimised to provide a maximum of absorption for hadronic showers before they reach the magnet coil. Therefore a brass/scintillator sampling technique is chosen. The HCAL covers an angular range of $|\eta| < 3.0$. Since there is still some hadronic leakage outside the magnet coil, an additional layer of scintillators is installed between magnet and return yoke so that the total number of hadronic interaction lengths, λ, is between 7 and 11. The hadron barrel ($|\eta| < 1.4$) is segmented in $\Delta\eta \times \Delta\phi = 0.087 \times 0.087$ large towers, while the endcap ($1.3 < |\eta| < 3.0$) is more coarse in ϕ with segmentations of 5° and 10°, respectively. In the forward region $3.0 < |\eta| < 5.0$, a steel and quartz-fibre calorimeter provides extended coverage for jet and missing E_T measurement.

The resolution of the reconstructed transverse energy of jets, E_T, is displayed in Fig. 5.15. It is about 15% for jets of $E_T = 100$ GeV and is below 10% for high energy jets. In the barrel region the resolution is parameterised by

$$\left(\frac{\sigma\left(E_T^{rec}/E_T^{MC}\right)}{\left\langle E_T^{rec}/E_T^{MC}\right\rangle}\right)^2 = \left(\frac{1.25}{\sqrt{E_T^{MC}}}\right)^2 + \left(\frac{5.6}{E_T^{MC}}\right)^2 + 0.033^2\,, \tag{5.13}$$

for jets reconstructed with an iterative cone algorithm with a cone size of $R = 0.5$. The calorimetric measurement of the missing transverse energy, E_T^{miss}, is important for searches for new particles and in the reconstruction of events which involve neutrinos in the final state. Figure 5.16 shows the corresponding expected resolution

Table 5.2 Comparison of performance parameters of the inner tracker systems of the ATLAS and CMS experiments [21]

Parameter	η	p_T[GeV]	ATLAS	CMS
Magnetic field B(T)	–	–	2	4
BR^2 ($T \cdot m^2$)	–	–	2.0–2.3	4.6–4.8
Silicon pixel channels	–	–	80×10^6	66×10^6
Pixel resolution in $R\phi$ (μm)	–	–	10	10
Pixel resolution in z/R (μm)	–	–	100	20
Silicon tracker channels	–	–	6.2×10^6	9.6×10^6
Tracker material thickness (X_0)	0	–	0.35	0.35
Tracker material thickness (X_0)	1.7	–	1.35	1.50
TRT tracker channels	–	–	0.35×10^6	–
$\mu^{\pm}$ reconstruction efficiency at	–	1	96.8%	97.0%
$\pi^{\pm}$ reconstruction efficiency at	–	1	84.0%	80.0%
$e^{\pm}$ reconstruction efficiency at	–	5	90.0%	85.0%
Momentum resolution at	0	1	1.3%	0.7%
	2.5	1	2.0%	2.0%
	0	100	3.8%	1.5%
	2.5	100	11%	7%
Transverse IP resolution (μm) at	0	1	75	90
	2.5	1	200	220
	0	1000	11	9
	2.5	1000	11	11
Longitudinal IP resolution (μm) at	0	1	150	125
	2.5	1	900	1060
	0	1000	90	22–42
	2.5	1000	190	40

for minimum-bias and soft QCD events at low-luminosity. The ATLAS and CMS hadronic calorimeter performances [21] are again compared in Table 5.3.

The most outer layers of the CMS detector are the muon systems. In the barrel section at $|\eta| < 1.2$ drift tube (DT) chambers are installed in four radial layers between $r = 4$ m and $r = 7$ m. The endcaps ($|\eta| < 2.4$) are equipped with three large disks of cathode strip chambers (CSC), which have a higher tolerance to neutron induced background and stand a higher particle rate and a higher magnetic field. In addition, resistive plate chambers (RPC) are used in both regions, which also operate at high rates. They provide a good time resolution needed for bunch crossing detection, but they have a reduced position resolution with respect to the DTs and CSCs. In the barrel, the DT and RPC systems provide up to 44 space points for muon track reconstruction. The muon chamber information is completed by the track measurement in the inner detector. In Fig. 5.16 the individual and combined momentum resolutions of the tracking systems are compared. At low transverse momentum, the resolution is mainly determined by the inner tracker, while the muon system contributes equally at high momenta. A similar behaviour is found in the forward region. For 1 TeV muons the momentum resolution is about 5 and 7% in barrel and endcaps, respectively. The corresponding angular resolutions in ϕ are 1 and 10 mrad. When compared to ATLAS, as summarised in Table 5.3, the

Table 5.3 Comparison of general performance parameters of the calorimeter and muon detectors of the ATLAS and CMS experiments [21]

Parameter	η	p_T[GeV]	ATLAS	CMS
Electromagnetic calorimeter				
Material	–	–	Pb/LAr	$PbWO_4$
Strips/Si-preshower thickness (X_0)	–	–	4.0–4.3	
Main sampling thickness (X_0)	–	–	16–20	25–26
Noise per cluster (MeV)	–	–	250	200–600
Resolution stochastic term	–	–	10–12%	3–5.5%
Resolution constant term	–	–	0.2–0.35%	0.5%
Hadronic calorimeter				
Material barrel	–	–	Fe/scint.	Brass/scint.
Material endcap	–	–	Cu/LAr	Brass/scint.
Material forward	–	–	Cu/LAr+W/LAr	Steel/quartz
Absorption lengths	–	–	9–13	7–14
$\pi^{\pm}$ Resolution stochastic term	–	–	55–85%	70%
$\pi^{\pm}$ Resolution constant term	–	–	1–2%	8%
Noise (GeV)	–	–	1.2–3.2	1
Muon spectrometer				
Measurement coverage	–	–	$\|\eta\| < 2.7$	$\|\eta\| < 2.4$
Trigger coverage	–	–	$\|\eta\| < 2.4$	$\|\eta\| < 2.1$
Bending power BL (m · T) at	0	–	3	16
	2.5	–	8	6
Combined (stand-alone) resolution at	0	10	1.4% (3.9%)	0.8% (8%)
	2	10	2.4% (6.4%)	2.0% (11%)
	0	100	2.6% (3.1%)	1.2% (9%)
	2	100	2.1% (3.1%)	1.7% (18%)
	0	1000	10.4% (10.5%)	4.5% (13%)
	2	1000	4.4% (4.6%)	7.0% (35%)

CMS stand-alone muon spectrometer has a less good resolution, which is however overcompensated in the combined reconstruction by the inner tracker.

The CMS trigger is a 3-layer system to select potentially interesting physics events which appear at rates much lower than the overwhelming inelastic *pp* collisions. At nominal bunch spacing and high luminosity, a reduction factor of 10^7 from about 1 GHz interaction rate to 100 Hz event recording frequency is needed. The trigger requirements of CMS and ATLAS are therefore very similar.

The first trigger layer of CMS is implemented in custom electronics while the high-level trigger (HLT) is software based. Only the calorimeters and the muon system provide trigger information to the hardware trigger. The ECAL trigger constructs trigger primitives from the sum of the transverse energy deposited in a calorimeter tower. The tower size in the barrel is 5×5 crystals, and in the endcap 5×5 so-called super-crystals. The HCAL trigger works similarly, with trigger towers that follow the HCAL geometry. The global trigger receives the trigger primitives and filters events that pass a pre-defined set of thresholds. The global trigger decision is prepared and sent back to the detector within 3.2 μs, which then starts the transfer of the buffered data. The level-1 muon trigger is based on DT, CSC and RPC measurements. Each subsystem provides a local trigger primitive

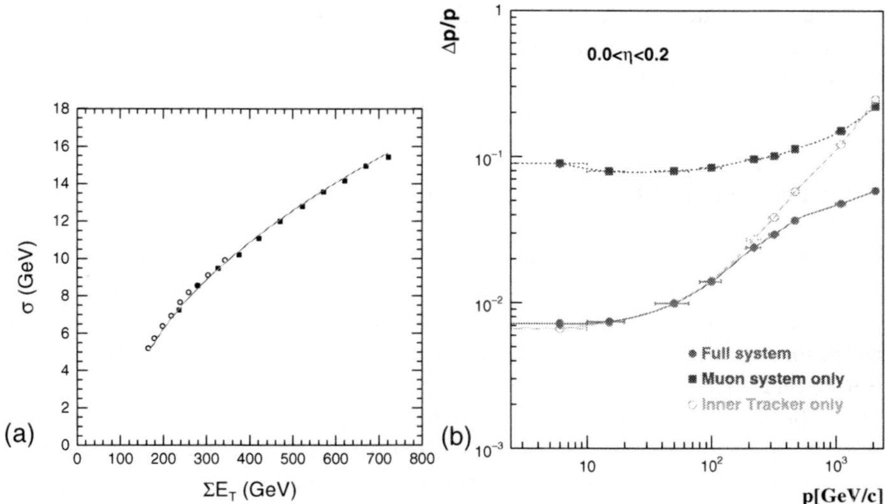

Fig. 5.16 (**a**) Missing transverse energy resolution as a function of the sum of the transverse energy, $\sum E_T$, for minimum-bias and soft QCD events at low luminosity [23]. (**b**) Resolution for muons in the central detector, reconstructed with the muon system only, the tracker only and the combined measurement [23]

with position, direction, bunch crossing and quality information. A track finding algorithm combines the primitives and assigns a p_T measurement to the track. The global muon trigger (GMT) selects the four best tracks according to p_T and quality and transmits it to the global trigger, which applies the required momentum and isolation thresholds. The global trigger can also combine different trigger primitives and has access to the calorimetric energy sums, $\sum E_T$ and E_T^{miss}. The level-1 accept rate is up to 100 kHz.

After a positive level-1 decision data are available in dual port memories to the DAQ system. The data blocks of the different detectors are accessible for the HLT processors which further analyse the events. The global strategy is to reject unwanted events as early as possible. Regional reconstruction is preferred in the early filtering stages because it can provide sufficient information to discard events quickly. In several virtual filter levels the complexity increases, eventually exploiting the completely reconstructed event, including full particle tracking. All HLT processors are running the same software code which maximises flexibility of the system. Eventually, after successful HLT filtering, event data are written to mass storage for offline analysis.

In summary, the ATLAS and CMS detectors will provide similar performances using different detection technologies. CMS has slight advantages in tracking charged particles in the inner detector and in the measurement of electron and photon energies. ATLAS, on the other hand, performs better in jet and energy flow determination, and also the stand-alone measurement of muon momenta in the muon chambers is more precise. This is however compensated in the combined reconstruction by the performance of the inner tracking system. The final performance will however also depend on the proper understanding of the sub-detectors and trigger systems, their efficient running, and the combined reconstruction of high p_T objects that will be optimised with data from pp collisions.

5.3 Prospects for the LHC Start-Up Phase

The sub-detectors and trigger systems of ATLAS and CMS were already operational at the LHC start in September 2008. The very first event in ATLAS is shown in Fig. 5.17. It is the interaction of a single proton beam hitting the upstream collimator and producing a shower of particles along the beam direction in the ATLAS detector. Before and after this event, several million of cosmic ray triggers were recorded for calibration and alignment. The sub-systems are completed, commissioned and, apart from temporary and system-specific faults related to normal operation, fully functional. The instantaneous LHC luminosity will increase with time such that the usage for measurements with the ATLAS and CMS detectors can be divided into different stages [22]. With 10–100 pb^{-1} of data collected in the very early phase (2010), detector calibration, trigger performance studies, trigger adjustment, and material studies will be performed. Known physics processes, like Drell-Yan Z- and W-Boson production, are used as standard candles for these tasks. Table 5.4 summarises some of the ATLAS performance goals for e/γ energy scale and uniformity, for jet energy scale, as well as for tracking and muon alignment.

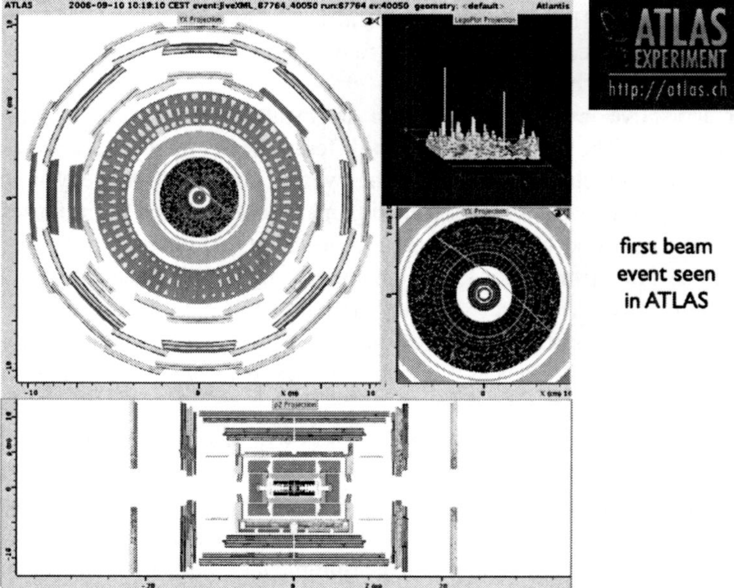

Fig. 5.17 First LHC beam event recorded by ATLAS

Table 5.4 Expected ATLAS calibration and alignment performance at the start of data taking and with first data samples

Quantity	Initial performance	Performance with data	Data samples
e/γ energy scale	2%	0.1%	$Z \to e^+e^-$, $J/\psi \to e^+e^-$
e/γ uniformity	1–2%	0.7%	$Z \to e^+e^-$
Jet energy scale	5–10%	1%	$W \to q\bar{q}$ in $t\bar{t}$, γ/Z + jets
Tracking alignment	10–50 μm	< 10 μm	Tracks, $Z \to \mu^+\mu^-$
Muon alignment	100–200 μm	30 μm	Inclusive muons, $Z \to \mu^+\mu^-$

In the subsequent phase, with up to about 1 fb^{-1} of data (expected in 2012), calibration and alignment will be further refined. Here, background processes for Higgs and SUSY searches need to be studied. Inclusive searches for SUSY particles, respectively their decays, will be possible in the low-mass SUSY parameter space with sensitivity to production cross-sections down to ≈ 0.5 fb.

Once the amount of well understood data goes beyond 1 fb^{-1}, the sensitivity extends to more rare processes, like the production SM and SUSY Higgs bosons as well as heavy new particles in the TeV range. Detailed information are collected in [23, 24] .

First LHC collision data will be used to verify and improve the calibration and alignment that has already been achieved with the corresponding dedicated hardware calibration systems. As an example, $Z \to e^+e^-$ events are planned to be used to inter-calibrate the different ATLAS calorimeter regions with a relative uniformity of 0.5% between regions of size $\Delta\eta \times \Delta\phi = 0.2 \times 0.4$. Together with the local uniformity obtained by cell-by-cell calibration, this is necessary to reduce the constant resolution term to below 0.7%. Figure 5.18 shows that only 100 pb^{-1} of data are needed to achieve this goal. The Z decay events serve also as a reference sample from which the absolute electromagnetic energy scale can be derived. Local energy scale factors are adjusted until the shape of the di-electron invariant mass distribution corresponds well to the Breit-Wigner line-shape folded with a resolution parameterisation, as expected for $Z \to e^+e^-$ production. Since the Z mass is known to 2.1 MeV from LEP measurements [6], the electron energies can be calibrated with high precision.

The $Z \to \ell^+\ell^-$ decays are also an ideal tool to measure lepton reconstruction, identification and trigger efficiencies, as well as resolutions directly from data [25]. The events are triggered and selected by requiring a high-p_T lepton to tag the event and a second object in an invariant mass interval close to the Z boson mass. This object is used as a probe to derive the various efficiencies. Figure 5.18 shows, as an example, the ATLAS muon identification efficiency, as it could be determined from 100 pb^{-1} of data. The relative background is small, less than 0.1%, and originating from $bb \to \mu\mu + X$ production, $W \to \mu\nu$ and $Z \to \tau\tau$ decays, as well as $t\bar{t}$ production. Similar measurements are foreseen for electrons and taus. The results of these studies will be compared to the estimated detector performance to eventually derive corresponding systematic uncertainties. They will be the base for searches for the Standard Model Higgs boson or for new physics beyond the Standard Model.

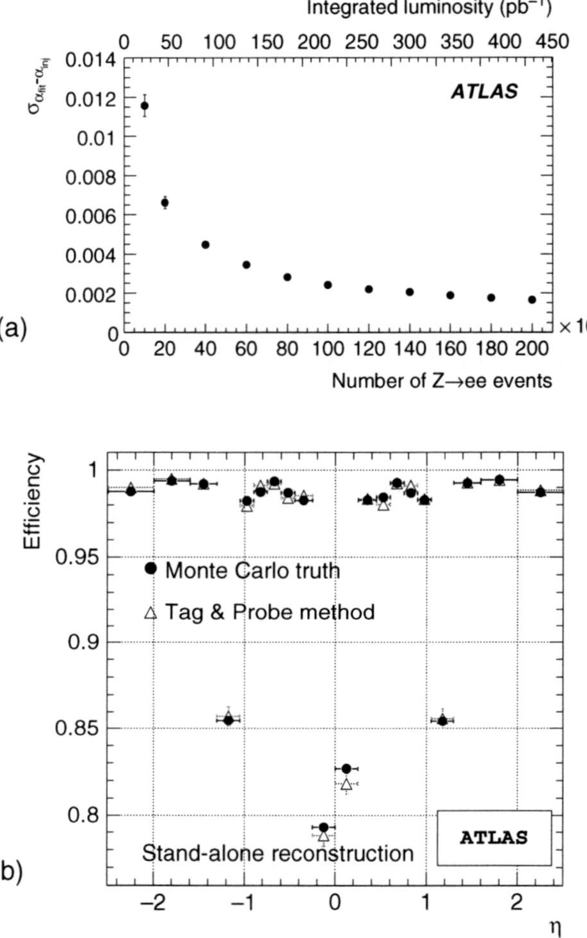

Fig. 5.18 (**a**) Constant energy resolution term from long-range non-uniformities in the calorimeter as planned to be measured from the line-shape of $Z \to e^+e^-$ events [24]. (**b**) Muon reconstruction efficiency determined from simulated $Z \to \mu^+\mu^-$ events and the corresponding background in 100 pb^{-1} of ATLAS data [24]. The result from the "tag&probe" method compares very well with the expectation directly derived from Monte Carlo information. The inefficiencies at $|\eta| \approx 0$ and $|\eta| \approx 1.2$ are due to the small gap between two muon barrel systems and the barrel-endcap transition region

References

1. The ATLAS Collaboration, G. Aad et al., J. Inst. **3** (2008) S08003.
2. The CMS Collaboration, S. Chatrchyan et al., J. Inst. **3** (2008) S08004.
3. L. Evans, Ph. Bryant, eds., *LHC Machine*, J. Inst. **3**, (2008) S08001.
4. The LHCb Collaboration, A. Augusto Alves Jr. et al., J. Inst. **3** (2008) S08005.
5. The ALICE Collaboration, K. Aamodt et al., J. Inst. **3** (2008) S08002.
6. The Particle Data Group, C. Amsler et al., Phys. Lett. **B 667** (2008) 1.

7. The Heavy Flavour Averaging Group, E. Barberio et al., *Averages of b-hadron and c-hadron Properties at the End of 2007*, arXiv:0808.1297.
8. The ATLAS Collaboration, *Detector and Physics Performance Technical Design Report*, CERN/LHCC/99-14/15, 1999.
9. Various aspects of QGP related physics can, e.g., be found in: J. Alam, S. Chattopadhyay, T. Nayak, B. Sinha and Y. P. Viyogi, eds., *Proceedings of Quark Matter 2008*, J. Phys. G: Nucl. Part. Phys. **35** (2008).
10. F. Karsch, E. Laermann and A. Peikert, Nucl. Phys. **B 605** (2001) 579.
11. U. Heinz and M. Jacob, *Evidence for a New State of Matter: An Assessment of the Results from the CERN Lead Beam Programme*, nucl-th/0002042v1, and references therein.
12. See for example: M. J. Tannenbaum, *Heavy Ion Physics at RHIC*, Int. J. Mod. Phys. **E 17** (2008) 771.
13. F. Carminati et al., J. Phys. G. Nucl. Part. Phys. **30** (2004) 1517; B. Alessando et al., J. Phys. G. Nucl. Part. Phys. **32** (2006) 1295.
14. T. Nayak, B. Sinha, *Search and study of Quark Gluon Plasma at the CERN-LHC*, arXiv:0904.3428v1; A. A. Isayev, *Physics of the ALICE experiment at the LHC*, arXiv:0810.4762v2.
15. M. H. Seymour, Nucl. Phys. **B 513** (1998) 269, hep-ph/9707338, and references therein.
16. S. Catani, Y. L. Dokshitzer, M. H. Seymour and B. R. Webber, Nucl. Phys. **B 406** (1993) 187 and refs. therein; S. D. Ellis and D. E. Soper, Phys. Rev. **D 48** (1993) 3160, hep-ph/9305266.
17. G. P. Salam, G. Soyez, JHEP 0705 (2007) 086, arXiv:0704.0292.
18. M. Cacciari, G. P. Salam, G. Soyez, JHEP **0804** (2008) 063, arXiv:0802.1189.
19. The ATLAS Collaboration, *ATLAS Forward Detectors for Measurement of Elastic Scattering and Luminosity*, ATLAS TDR 018, CERN/LHCC 2008-04.
20. S. Ask et al., Nucl. Inst. Meth. **A 568** (2006) 588; F. Anghinolfi et al., J. Inst. **2** (2007) P07004.
21. D. Froidevaux and P. Sphicas, Ann. Rev. Nucl. Part. Sci. **56** (2006) 375.
22. See for example: A. Straessner, *Proceedings Status of ATLAS and Expectations for First Physics*, Advanced Studies Institute on Symmetry and Spin, Prague, 2008, ATLAS Proceedings, ATL-GEN-PROC-2009-001.
23. The CMS Collaboration, G.L. Bayatian et al., CMS Physics Technical Design Report, Vol. II: Physics Performance, CERN/LHCC 2006-021.
24. The ATLAS Collaboration, G. Aad et al., *Expected Performance of the ATLAS Experiment Detector, Trigger, Physics*, CERN-OPEN-2008-020, arXiv:0901.0512.
25. A. Straessner and M. Schott, *A New Tool For MeasuringDetector Performance in ATLAS*, Proceedings of the CHEP09 Conference, submitted to Journal of Physics.

Chapter 6
Expectations for Electroweak Measurements at the LHC

At the LHC, the large amount of luminosity expected will allow a determination of the masses of the W boson and of the top quark with even greater precision than present day measurements at LEP and the Tevatron. This is a very challenging task. In particular, the experimental and systematic uncertainties need to be under control. There is also the opportunity to improve the knowledge on the weak mixing angle and on triple gauge boson couplings. This will further constrain the theoretical models and will probe the consistency of the Standard Model in finer detail. This chapter focuses therefore on the prospects for improved electroweak measurements by the LHC experiments.

6.1 W and Z Boson Production

The production of W and Z bosons is a process with high event rates at the LHC, with NNLO cross-sections of 20.5 nb [1] for $W \to e\nu_e, \mu\nu_\mu$ final states and about 10 times smaller, 2.02 nb [1], for $Z \to \mathrm{e^+e^-}, \mu^+\mu^-$ production at $\sqrt{s} = 14$ TeV. The theoretical predictions performed at NNLO have a rather high precision of about 1% and the cross-section measurements provide stringent tests of QCD. In addition, differential cross-sections like the rapidity distribution of W and Z bosons, $d\sigma/dy$, are sensitive to parton distribution functions (PDFs) of the proton (compare Sect. 1.7 and References [10, 39]).

In the initial LHC phase, at low luminosities of 10^{31} cm^{-2}s^{-1}, the trigger thresholds for W $\to \ell\nu$ and Z $\to \ell^+\ell^-$ production can be rather low. Single electron and muon signatures are planned to be accepted by the first level trigger at $p_T > 10$ GeV. Electron and muon pairs are required to pass at least $p_T > 5$ GeV and $p_T > 4$ GeV, respectively. These thresholds are approximately doubled at 10^{33} cm^{-2}s^{-1} and will be further adapted for nominal LHC running at 10^{34} cm^{-2}s^{-1}. The selection of $W \to e\nu$ events, as prepared by the ATLAS collaboration, is subsequently asking for a well identified electron reconstructed from an electromagnetic cluster in the calorimeter matched in angle to a track, such that $E_T > 25$ GeV and η values within the fiducial volume [2]. The missing transverse energy due to the neutrino has to pass $E_T^{miss} > 25$ GeV and the transverse mass of the lepton-neutrino system, m_T^W, must be larger than 40 GeV.

A. Straessner, *Electroweak Physics at LEP and LHC*, STMP 235, 167–187,
DOI 10.1007/978-3-642-05169-2_6, © Springer-Verlag Berlin Heidelberg 2010

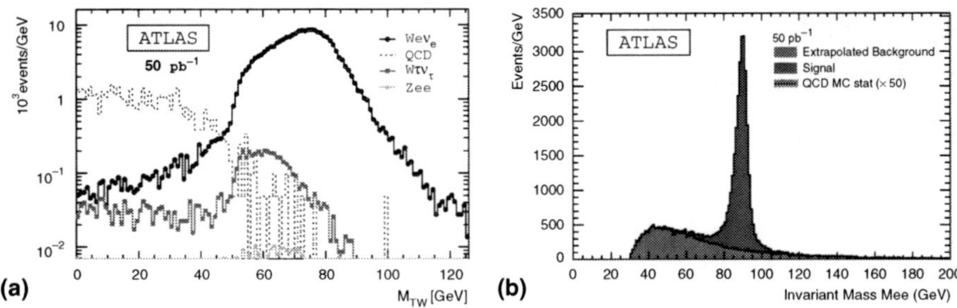

Fig. 6.1 (**a**) Simulated transverse mass distribution in the $W \to e\nu_e$ channel for an integrated luminosity of $50\,\mathrm{pb}^{-1}$ [2]. (**b**) Invariant di-electron mass measured in $Z \to e^+e^-$ events and the corresponding background, again for $50\,\mathrm{pb}^{-1}$ [2]

In $50\,\mathrm{pb}^{-1}$ of data, 220 thousand signal events are expected. Main backgrounds are QCD jet production, followed by $W \to \tau\nu$ and $Z \to e^+e^-$ events, summing up to about 10% of the signal expectation. The transverse mass distribution is shown in Fig. 6.1 [2].

Since QCD jet production is the primary background and has at the same time relatively large theoretical uncertainties, a data driven method is developed for its estimation. Especially the modelling of the E_T^{miss} background distribution before the final cut is important. A $\gamma + jets$ event sample, very similar to the signal process, is selected by requiring that no charged tracks are pointing to the electromagnetic cluster instead of having an angular match. Thus, the control sample has similar kinematics to the signal but a priori no missing energy like the background. Therefore, the background E_T^{miss} spectrum can be derived from these events and systematic uncertainties are reduced. Table 6.1 summarises the event numbers, acceptances, A, and efficiencies, ε, as well as the prediction for the cross-section measurement in $50\,\mathrm{pb}^{-1}$ of data.

The $W \to \mu\nu$ selection follows a similar strategy. However, the jet backgrounds are dominated by $bb \to \mu X$ events which can be rejected by requiring muon isolation from hadronic activity and impact parameter cuts. A good understanding of the detector and the underlying event is necessary. Luminosity uncertainties will be significant during initial running, but can be removed by taking ratios of cross-sections σ_W/σ_Z. The situation will improve once the absolute luminosity calibration

Table 6.1 Expected number of signal and background events, N and B, overall selection efficiencies, $A \times \varepsilon$, and cross-section measurements, σ, together with their uncertainties, for an integrated luminosity of $50\,\mathrm{pb}^{-1}$. The uncertainty on N is statistical, the other sources are systematic [2]. An overall luminosity uncertainty is not included

Process	$N(\times 10^4)$	$B(\times 10^4)$	$A \times \varepsilon$	$\delta A/A$	$\delta\varepsilon/\varepsilon$	σ (pb)
$W \to e\nu$	22.67 ± 0.04	0.61 ± 0.92	0.215	0.023	0.02	$20520 \pm 40 \pm 1060$
$W \to \mu\nu$	30.04 ± 0.05	2.01 ± 0.12	0.273	0.023	0.02	$20530 \pm 40 \pm 630$
$Z \to e^+e^-$	2.71 ± 0.02	0.23 ± 0.04	0.246	0.023	0.03	$2016 \pm 16 \pm 83$
$Z \to \mu^+\mu^-$	2.57 ± 0.02	0.010 ± 0.002	0.254	0.023	0.03	$2016 \pm 16 \pm 76$

with the ALFA detector [3] will be available, for which dedicated runs are foreseen. The absolute luminosity calibration will then be projected to normal ATLAS running, using e.g. the relative measurement of the LUCID system.

The measurement of single gauge boson production thus represents a first test of the Standard Model predictions, testing both the electroweak and QCD part of the theory. Verifying the latter in the high energy hadronic environment is especially interesting and the basis for many other measurements and searches at the LHC. Controlling and measuring PDFs, as one aspect of understanding the hadronic part of the interactions, will be an important activity.

Insights into beyond leading-order QCD jet production can be learnt by selecting explicitly $Z \to \mu^+\mu^- + jets$ signatures from the $Z \to \mu^+\mu^-$ sample (and similarly for $Z \to \mathrm{e}^+\mathrm{e}^-$). This is interesting by itself and necessary to understand backgrounds to new particle searches. In an ATLAS study [2], di-muon events in the mass range 81 GeV $< m_{ee} <$ 101 GeV with isolated muons of high p_T are selected. Jets are identified with a minimal angular distance $\Delta R > 0.4$ with respect to the muon, a minimal transverse momentum, $p_T > 40$ GeV, and a pseudo-rapidity range in the central part of the detector, $\eta < 3.0$. The purity for $Z \to \mu^+\mu^-$ with additional 1-, 2- and 3-jets is found to be rather high, but to decrease with the jet multiplicity from about 96 to 90%.

The spectrum of the jets is determined and then corrected back to Monte Carlo generator level, using the ALPGEN [4] calculation as a reference. Figure 6.2 compares the different predictions at parton level. The actual comparison is done on hadron-level but reveals similar features: the LO prediction of the PYTHIA program differs from the NLO MCFM [5] and ALPGEN calculations, especially in the high jet p_T region. Fig. 6.2 shows also the expected precision for different systematics due to the jet energy scale. Initially, this scale will not be known better than 10% (with $\mathcal{L} = 1\,\mathrm{fb}^{-1}$). But with more data, a 5% precision is expected to be reached, providing sensitivity to LO vs. NLO differences.

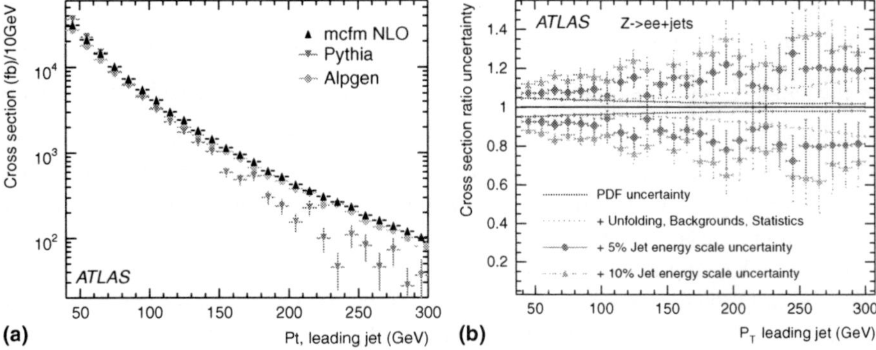

Fig. 6.2 (**a**) Parton level comparison of the p_T of the leading jet in $Z \to \mu^+\mu^- + jets$ Monte Carlo samples for LO and NLO predictions [2]. (**b**) Systematic uncertainty on hadron level p_T of the leading jet in $Z \to \mathrm{e}^+\mathrm{e}^- + jets$ events for $\mathcal{L} = 1\,\mathrm{fb}^{-1}$ [2]. If the dominant jet energy scale uncertainty can be reduced below 10%, sensitivity to NLO predictions is possible

6.2 W Mass Measurement at the LHC

The techniques foreseen to determine the mass of the W boson at the LHC are very similar to those applied at the Tevatron. The data samples consist of leptonically decaying W boson with eν and $\mu\nu$ final states. Events with high-p_T leptons, typically above 20 GeV, and missing energy of more than 20 GeV due to the unmeasured neutrino are selected. Activity from QCD jets is suppressed by limiting the hadronic recoil to the reconstructed W boson to 30 GeV. This retains about 40 million W decays per experiment in 10 fb^{-1} of data with efficiencies of 20% in each decay channel $W \rightarrow e\nu$ and $W \rightarrow \mu\nu$ [6]. With this number of signal events the statistical precision is only a subordinate uncertainty. Controlling the different sources of systematics is therefore the main task in this measurement. The goal is to push the experimental precision close to the current theoretical uncertainty of 4 MeV [7], calculated at electroweak two-loop order (and for $M_{\mathrm{H}} < 300$ GeV).

In general, many effects can be determined from the closely related single gauge boson channels, $Z \rightarrow e^+e^-$ and $Z \rightarrow \mu^+\mu^-$. It could be shown [6] that from Z control samples, the uncertainties on the W rapidity spectrum, the W p_T spectrum, as well as the lepton energy scale and resolution as a function of lepton energy can be derived. Like for the Tevatron measurements, the $Z \rightarrow \ell^+\ell^-$ samples can be well calibrated by applying the nominal Z mass [8] as an external constraint. Similarly, the lepton energy scale can be determined and calibrated for the lower lying di-lepton resonances, like J/ψ and Υ. CMS had proposed a method to remove one of the leptons of the Z decay and adjust the p_T, respectively E_T, and the transverse mass spectra such that the resulting distributions fit the measured ones in the corresponding $W \rightarrow \ell\nu$ channel. The adjustment is done with the following approximation:

$$\left.\frac{d\sigma^W}{dp_T}\right|_{pred} = \frac{M_Z}{M_W}\left.\left(\frac{d\sigma^W}{M_W dp_T} \Big/ \frac{d\sigma^Z}{M_Z dp_T}\right)\right|_{theo} \left.\left\{\frac{d\sigma^W}{dp_T}\left(p_T^Z - \frac{M_Z}{M_W} p_T^W\right)\right\}\right|_{meas}, \tag{6.1}$$

where M_{W} is a free fit parameter and many theoretical uncertainties are absorbed in the ratio, $\left.\left(\frac{d\sigma^W}{M_W dp_T} / \frac{d\sigma^Z}{M_Z dp_T}\right)\right|_{theo}$, which is used as theoretical input. This is illustrated in Fig. 6.3. Since then, more refined methods like the template method, also applied in top and W mass analyses at Tevatron and elsewhere, are used. They yield a very high statistical precision of about 2 MeV per channel in 10 fb^{-1} of data. The method relies on Monte Carlo samples of signal events which are reweighted to different underlying masses using theoretical matrix element predictions which are adjusted to fit the data spectra of transverse lepton momentum and transverse mass.

A detailed study of the different sources of systematics [6] shows that many uncertainties can be controlled by studying event samples of leptonically decaying Z bosons. The size of these samples is a factor 10 smaller than the number of W boson decays in the respective channels (compare Table 6.1). The leptonic Z final state can however be fully reconstructed and the kinematic properties of the Z can

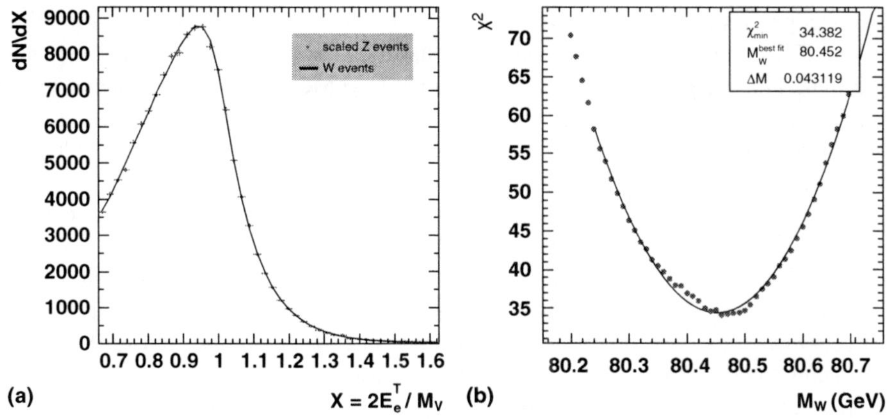

Fig. 6.3 (**a**) Comparison of the rescaled E_T spectrum of electrons in $Z \rightarrow e^+e^-$ events 1 fb^{-1} [9]. (**b**) The corresponding χ^2 distribution of the W mass derived from this spectrum

be determined. Assuming that each leptonic energy, E_i ($i = 1, 2$), scales with a factor α_i the reconstructed Z mass, m_{ij}, varies as

$$\alpha_{ij} m_{ij} = \sqrt{2\alpha_i E_i \alpha_j E_j (1 - \cos \Delta\phi)} \tag{6.2}$$

$$\alpha_{ij} \approx \frac{1}{2}(\alpha_i + \alpha_j), \tag{6.3}$$

with the angle between the leptons, $\Delta\phi$, and an approximative expression for the mass scale-factor, α_{ij}. In this way, the measured Z mass spectrum is adjusted to follow the theoretical expectations. The energy scale and resolution parameters are then derived as a function of the lepton energy itself with high precision. The impact of energy scale and resolution on the W mass are thus reduced to 4 MeV and 1 MeV, respectively, for both electrons and muons.

Furthermore, energy and pseudo-rapidity dependent trigger, reconstruction and identification efficiencies, as well as the recoil energy scale is extracted from Z decay events. Uncertainties on efficiencies may alter the p_T and m_T spectra, and influence eventually M_W by 4.5 MeV and 1 MeV in the electron and muon channel, respectively. When studying the recoil system, the leptons are properly removed from the event to simulate and then measure missing energy signatures, very similar to the neutrino in $W \rightarrow \ell\nu$ decays. The extraction of M_W from the transverse mass has a corresponding systematic uncertainty of 5 MeV. These estimates are performed based on Monte Carlo simulations and exploit experience from the Tevatron precision measurements. However, the detailed understanding of the detectors of ATLAS and CMS is primordial to achieve such a performance.

The modelling of the W production mechanism also must be understood, in particular the p_T and rapidity distribution of the W bosons. Here, the Z decay studies and Drell-Yan di-lepton events help to constrain these. Extrapolating the current knowledge on PDFs, e.g. by analysing the systematically varied PDF sets provided by the CTEQ group [10], and assuming an significant improvement, the corresponding uncertainty on M_W can be reduced to below 3 MeV level. Modelling of QED photon radiation from final state leptons contributes with only 1 MeV, expecting a similar level of understanding of the photon spectra and their influence on the lepton reconstruction as for the LEP Z mass measurements.

As regards backgrounds, their contribution to the selected event sample is only small, mainly dominated by irreducible $W \rightarrow \tau\nu \rightarrow e\nu\nu$ and $W \rightarrow \tau\nu \rightarrow \mu\nu\nu$ events with the same signature as the signal process. An uncertainty of about 2 MeV is attributed to systematic variations of the background level and its spectral shape. Effects from the underlying event, in-time and out-of-time pile-up, and variations of the beam crossing angle will influence the W mass by only 1 MeV.

In total, the systematic uncertainties sum up to 6–7 MeV in the muon channel and to 7–8 MeV in the electron channel for each of the mass measurements from the p_T and m_T spectra. Taking correlations into account this yields an accuracy of 7 MeV in both channels [6]. As mentioned above, this estimate assumes $10\,\mathrm{fb}^{-1}$ of well understood data. Only data taken with close-to-ideal detector conditions and under well controlled beam conditions with negligible background will enter the W mass analysis.

Since both experiments, ATLAS and CMS, will independently measure M_W, a combined uncertainty (including correlations) very close to the current theoretical precision of 5 MeV will therefore be possible, improving the current world average by a factor of 5. This will then match well with the uncertainty on M_Z of 2.1 MeV. Theoretical models are thus tested in future also in the W mass sector at the electroweak two-loop level, and possibly beyond.

6.3 Top Physics and Determination of the Top Quark Mass

The most prominent Standard Model process at the LHC is the top quark production, making the LHC a top factory. About 83,000 top pairs are expected in $100\,\mathrm{pb}^{-1}$ [11] at a centre-of-mass energy of 14 TeV. They are produced through gluon fusion diagrams $gg \rightarrow t\bar{t}$ (90%) and quark annihilation $qq \rightarrow g \rightarrow t\bar{t}$ (10%). The cross-section depends on the exact value of the top quark mass, m_t, but can be calculated at NLO order including NLL soft gluon resummation. The renormalisation scale uncertainty is however non-negligible, in the order of 10% when the scale is varied by a factor of two [12]. When including NNLO calculations and using the minimal subtraction ($\overline{\mathrm{MS}}$) renormalisation scheme to define the top quark mass, $m_t(\mu_r)$, an improved theoretical uncertainty of below 5% is achieved [13].

The top quarks decay practically exclusively to W+b since the CKM matrix element V_{tb} is close to unity. The $t\bar{t}$ events are therefore measured in three topologies according to the W decay final states: fully hadronic (46.2%), semi-leptonic (43.5%) and fully leptonic (10.3%). The trigger systems of ATLAS and CMS identify those events by multiple signatures: high-p_T jets, isolated high-p_T electrons and muons in the leptonic channels, and multi-jets in the fully hadronic channel.

Typical efficiencies normalised to the total event rate are in the order of 50–60% for the lepton triggers with $p_T > 20 - 25$ GeV, nearly 100% for low threshold jet triggers with $p_T > 20$ GeV and about 10% for multi-jet triggers. Especially in the semi-leptonic final state there is a large redundancy.

The event selection in the single lepton channel requires a high p_T lepton of 20 GeV, $E_T^{miss} > 20$ GeV, four jets of $p_T > 20$ GeV with three jets passing $p_T > 40$ GeV. This results in combined trigger and selection efficiencies of 18% in the electron and 24% in the muon channel. Furthermore, additional kinematic cuts can be applied like W mass constraints and a top-mass window, as well as b-tagging. The latter is however considered as not applicable in very early data since it requires a thorough understanding of the ATLAS tracking. Without asking for a b-tag, the electron analysis expects a signal-to-background ratio of $N_S/N_B = 561/96$ events in $100\,\text{pb}^{-1}$ and the muon analysis is expecting a ratio of $N_S/N_B = 755/143$. Events from $W + jet$ production represent the main background. The cross-section is extracted from a likelihood fit to the three-jet mass spectrum, as shown in Fig. 6.4, which yields a relative precision of $\Delta\sigma/\sigma = (7(\text{stat.}) \pm 15(\text{syst.}) \pm 3(\text{PDF}) \pm 4(\text{lumi.}))\%$ for both channels combined. The systematics are dominated by initial and final state radiation (ISR/FSR) of gluons and photons, as well as the shape of the fit function used.

In the di-lepton channel, the typical signature are two high p_T leptons, E_T^{miss} due to two neutrinos which escape detection and two high p_T b-jets. Combining ee, $e\mu$ and $\mu\mu$ channels, the signal to background ratio is $N_S/N_B = 987/228$. Here, the leptonic decays of Z and W bosons are dominating the background rate. The expected precision in $100\,\text{pb}^{-1}$ is $\Delta\sigma/\sigma = \left(4(\text{stat.})^{+5}_{-2}(\text{syst.}) \pm 2(\text{PDF}) \pm 5(\text{lumi.})\right)\%$ using a simple event counting method [2]. In this case, the jet energy scale is expected to be the main source of systematic uncertainties.

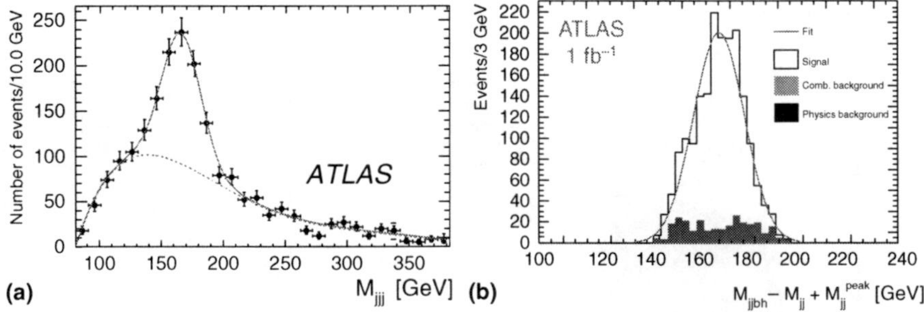

Fig. 6.4 (**a**) Reconstructed top mass in the decay $t \to Wb \to jjb$ in semi-leptonic $t\bar{t}$ events for $\mathcal{L} = 100\,\text{pb}^{-1}$ [2]. From this distribution the $t\bar{t}$ cross-section is determined by fitting a parameterised signal and background function to the simulated data. (**b**) A high purity $t\bar{t}$ sample is used to measure the top quark mass [2]. An increased width of the jjb mass due to the light jet energy scale is corrected by adding the jj mass peak-value, M_{jj}^{peak}, instead of the actual jj mass, M_{jj}, in each event. This corrects the jet energy scale uncertainty to first order. The statistical uncertainty on m_t in $1\,\text{fb}^{-1}$ obtained in this case is 0.3 GeV

The top events themselves, and in particular the hadronic W decays, can actually be explored to calibrate the jet energy scale in data. Knowing the W mass value, the invariant jj mass spectrum can be adjusted to the expectations. Iterative energy rescaling and template methods are used. As an example, in 4,000 semi-leptonic $\mathrm{t\bar{t}}$ events from 1 fb^{-1} of simulated data, an overall scale factor of $K = M_{\mathrm{W}}^{PDG}/M_{jj} = 1.014 \pm 0.003$ is achieved, reproducing well the expected value of the absolute jet energy scale of $K_{exp} = E_{parton}/E_{jet} = 1.014 \pm 0.002$ for the specific sample analysed.

The jet energy scale is also the main uncertainty in the determination of the top quark mass. A first study for this measurement was performed in the semi-leptonic channel. The purity of the $\mathrm{t\bar{t}}$ sample is increased by additional kinematic constraints, e.g., on the reconstructed hadronic W mass, the b-quark energy, E_b^*, and the difference of b and W energies, $E_W^* - E_b^*$, in the top rest frame. Eventually, top, W and b purities of (86.4±0.9)%, (86.9±0.9)% and (94.0±0.6)% are reached. The selection efficiency is (0.57 ± 0.05)%. From the top mass spectrum, shown in Fig. 6.4, m_{t} is derived with a very good precision of 0.3 GeV, assuming $\mathcal{L} = 1\,\mathrm{fb}^{-1}$, with practically no bias. This means that the systematic uncertainties dominate, as there are: b-jet energy scale with 0.7 GeV per %, light jet energy scale 0.2 GeV per %, and ISR/FSR systematics of 0.4 GeV. The ultimate goal is therefore the reduction of the jet energy scale uncertainties to at least 1%, which is one of the main challenges in this measurement [2, 9]. The Tevatron measurements of m_{t} (Sect. 4.2) already push forward into these regions of systematic uncertainties. An improvement by the LHC experiments of the top quark mass will therefore require a detailed understanding also of the modelling of the underlying event, of measurements and modelling of the much larger pile-up, and of a measurement and control of QCD effects like Colour Reconnection. Here, the large data samples which will be collected by the LHC experiments may give a handle to possibly select a set of $t\bar{t}$ events in which Colour Reconnection can either be measured or in which this effect is suppressed. Then the LHC will eventually be able to achieve a precision of the top quarks mass at or even below the 1 GeV level.

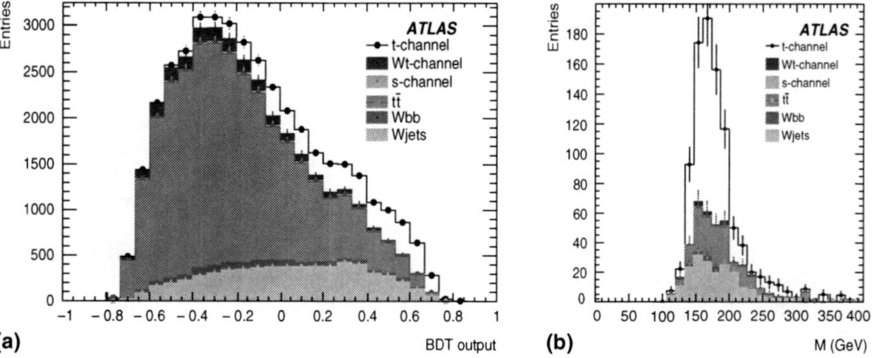

Fig. 6.5 (**a**) Boosted decision tree (BDT) output for single-top signal and background [2]. (**b**) Leptonic top quark mass distribution applying cut on the BDT output at 0.6 [2]

An alternative method to the direct determination of the top quark mass from the reconstruction of the top decay is the extraction of m_t from the top-pair production cross-section, $\sigma_{t\bar{t}}$. This method is already applied at the Tevatron [18] with uncertainties of 5.5–6 GeV on the top quark mass. The dependence of $\sigma_{t\bar{t}}$ on m_t, needed for this measurement, was calculated at NLO [14, 15] and NNLO accuracy [16, 17]. At this level of theoretical accuracy, it turns out that the definition of m_t as the pole mass is not safe with respect to the order of perturbation theory that is applied. A running mass with renormalisation scale, $m(\mu_r)$, is theoretically better defined since the top is a coloured object. QCD confinement prevents to extract the pole mass in the top-quark channel. The pole mass is usually converted into $m(\mu_r)$ using the following NNLO relation [16]:

$$m_t = m(\mu_r)\left(1 + \frac{\alpha_s(\mu_r)}{\pi}d_1 + \left(\frac{\alpha_s(\mu_r)}{\pi}\right)^2 d_2\right), \tag{6.4}$$

with μ_r-dependent coefficients d_1 and d_2. If the cross-section predictions are compared to DØ measurements at the Tevatron of $\sigma_{t\bar{t}} = 8.18^{+0.96}_{-0.87}$ [18] one can extract a pole mass of 168.2 ± 3.6 GeV [16] (The more precise CDF result of $\sigma_{t\bar{t}} = 7.0 \pm 0.6$ [19] is not used in the analysis). The $\overline{\text{MS}}$ mass, $m(m)$, yields a lower value of 160 ± 3.3 GeV, however, much more stable against contributions from higher order effects. These arguments need to be taken into account also for a more precise determination of m_t from the top decay spectrum. Currently, improved LO and NLO Monte Carlo predictions are used to derive the pole top mass using template or other unfolding methods [20]. The renormalisation scheme must therefore be much better studied and defined if top masses from top decay measurements and from top-pair cross-section data shall be compared or even combined. This is important in particular for the presumably more precise measurements at the LHC.

In proton–proton collisions, top quarks are not only produced in pairs but also in single-top processes, where the electroweak t-channel production, $qg \to q' + t\bar{b}$ and $qb \to q't$, dominates with $\sigma_t = 246 \pm 12$ pb [21]. The Wt-channel, $gb \to b \to Wt$, is contributing with 66 ± 2 pb [22] and the s-channel, $q\bar{q}' \to W \to t\bar{b}$, with 11 ± 1 pb [21]. Single-top production is especially interesting because the cross-section is directly proportional to the CKM matrix element $|V_{tb}|^2$. The backgrounds from $t\bar{t}$, $W + bb$ and $W + jets$ are very difficult to reject. Multivariate analyses are therefore applied using variables like b-jet p_T and η, ΔR between jets and leptons, m_T^W, etc. Figure 6.5 shows the output of a so-called boosted decision tree (BDT) analysis [23] and the top mass spectrum for high purity events. Assuming $\mathcal{L} = 1$ fb^{-1}, a relative precision in the cross-section measurement of $\Delta\sigma/\sigma = (5.6(\text{stat.}) \pm 22(\text{syst.}))\%$ can be achieved. Systematic effects from b-tagging, jet energy scale, and PDFs contribute the most to the total uncertainty. Translated into a measurement of $|V_{tb}|$ one can derive $\Delta|V_{tb}|/|V_{tb}| = (11(\text{stat.+syst.}) \pm 4(\text{theo.}))\%$, where the theory uncertainty takes strong scale and PDF dependencies into account. The estimated precision is very much compatible with the one obtained in about 3 fb^{-1} by the Tevatron experiments [24, 25], which yield, in case of the CDF measurement, $|V_{tb}| = 0.91 \pm 0.11(\text{exp.}) \pm 0.07(\text{theo.})$. The s- and Wt-channels are studied as well for the early LHC scenario, but a few fb^{-1} will be needed to establish a signal with more than 3 standard deviations.

6.4 Measurement of Triple Gauge Boson Couplings

Similar to e^+e^- colliders, the gauge bosons W, Z and γ can be produced in pairs at hadron colliders like the LHC. Possible final states include all di-boson combinations, from which the most interesting are WW, ZZ, WZ, Wγ, and Zγ, since they may involve triple gauge boson couplings (TGCs). The generic lowest order Feynman diagrams are shown in Fig. 6.6.

At a centre-of-mass energy of 14 TeV, the W-pair, WZ and Z-pair production cross-sections amount to 111.6, 47.8 and 14.8 pb [26], respectively, where the first also takes finite width effects of the intermediate bosons into account. These cross-sections are a factor 10 higher than at the Tevatron at $\sqrt{s} = 1.96$ TeV. All di-boson processes have been observed at the Tevatron and cross-sections are measured, including ZZ production which has the lowest rate. The photons in Wγ and Zγ production are typically required to have high transverse energy with respect to the beam directions and large angle to the final state fermions from W and Z decays, to separate the di-boson process process from QED photon radiation. Asking for $E_T^{\gamma} > 7$ GeV and $\Delta R(f, \gamma) > 0.7$ yields cross-sections of 451 pb [27] for Wγ and 219 pb [28] for Zγ production.

Deviations from the Standard Model predictions due to anomalous TGCs are described with the same effective Lagrangian used in TGC analysis at LEP. The charged couplings terms, with couplings g_1^Z, κ_Z, κ_γ, λ_γ and λ_Z, are given in Eq. (3.47). The coupling structure of the neutral sector, detailed in Eq. (3.37), contains two anomalous couplings, f_4 and f_5. The terms beyond the Standard Model have the common characteristic to increase with momentum scale of the process. The resulting cross-section will therefore violate unitarity at some high scale Λ^2, where new, yet unknown physics must set in to regulate this behaviour. To incorporate this potential extension of the Standard Model, a cut-off term is introduced in the anomalous parts of the couplings:

$$\Delta g(\hat{s}) = \frac{\Delta g}{\left(1 + \frac{\hat{s}}{\Lambda^2}\right)^n}, \tag{6.5}$$

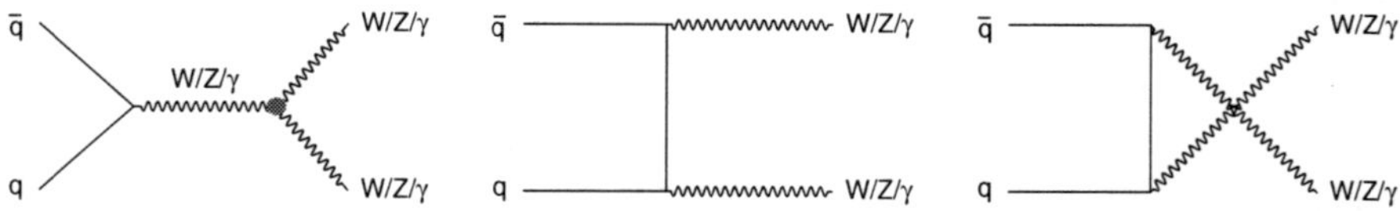

Fig. 6.6 Lowest order Feynman diagrams for di-boson production at hadron colliders. The s-channel diagram (*left*) involves a triple gauge boson vertex, which may differ from the Standard Model coupling structure. In the Standard Model, only WWZ and WWγ vertices are allowed at tree level

where g stands generically for one of the couplings listed above, and Δg describes its deviation from the Standard Model value. The cut-off scales with the ratio of four-momentum transfer $\hat{s}$ to Λ^2. For charged couplings the power n is chosen to be equal to 2, while for neutral couplings n equals 3. This compensates different powers of momenta in the interaction terms.

The various di-boson final states have different sensitivities to the charged couplings due to the given coupling structure. The $\Delta\kappa_\gamma$ and $\Delta\kappa_Z$ terms in WW production scale with di-boson mass squared, $\hat{s}$, while it is only proportional to $\sqrt{\hat{s}}$ in WZ and Wγ production. Using the analogue argument, the WZ final state is more sensitive to Δg_1^Z than WW production. In case of λ_γ and λ_Z, the corresponding terms scale with $\hat{s}$ in all WW, WZ and Wγ channels. For this reason, the LHC experiments are expected to have an enhanced sensitivity to anomalous TGCs due to the high centre-of-mass energy of 14 TeV.

For numerical evaluation of the LHC performance, the Monte Carlo programs BosoMC [27] and BHO [28] are used. These are accurate to NLO, and compare well to standard generators [2] if anomalous couplings are turned off. In case of a variation of neutral couplings in the ZZ channel, a p_T dependent K-factor is applied to reach NLO accuracy.

The selection of the di-boson events is concentrating on the final states where W and Z bosons decay leptonically: W $\rightarrow \ell\nu$ and Z $\rightarrow \ell^+\ell^-$, where the leptons are either muons or electrons. The WZ $\rightarrow \ell\nu\ell^+\ell^-$ events are characterised by three high p_T leptons, two of the same flavour, and missing transverse energy due to the neutrino. The main backgrounds consist of ZZ $\rightarrow \ell^+\ell^-\ell^+\ell^-$ events, where one lepton escapes undetected, of Z $+ jet/\gamma$, where the jet or photon is misidentified as a lepton, and of top-pair events with missing energy and three final state leptons coming from W and b quark decays. Strict cuts on a di-lepton mass window and a transverse mass formed with the third lepton and missing transverse energy consistent with Z and W mass, respectively, are applied. Events containing activity from hadronic jets are furthermore suppressed. In this way, a signal-to-background ratio of 7 with a number of signal events of about 50 in 1 fb^{-1} of data are expected. This can be further improved by using multi-variate analysis (MVA) techniques [2].

Production of Wγ events with sensitivity to WWγ couplings are enhanced by the kinematic cuts mentioned above, which define the signal process. The selection follows mainly the standard W selection with additional criteria to identify isolated photons. Background from final state photon radiation and misidentified photons is however difficult to suppress. With optimised MVA analyses it is expected to find about 1,600 signal and 1,200 background events in 1 fb^{-1} of data in the W$\gamma \rightarrow e\nu\gamma$ channel. The selection of W$\gamma \rightarrow \mu\nu\gamma$ events is about 50% more efficient, similar to the inclusive W boson selection, listed in Table 6.1.

Final states with W bosons are completed by the WW $\rightarrow \ell\nu\ell\nu$ channel. Two isolated, high-p_T leptons above 25 GeV with opposite sign and measured in the central detector, $|\eta| < 2.5$, are require in the event selection. A jet veto reduces top-pair background, and a minimal $E_T^{miss} > 50$ GeV suppressed Drell-Yann production, Z/$\gamma^* \rightarrow \ell^+\ell^-$. When the WW cross-section is measured, a cut on the angle between the two leptons, $\Delta\phi_{\ell\ell} < 2$, improves the purity. However, for TGC extraction it is more advantageous to require a minimal separation angle between the p_T^{miss} vector and the momentum of the di-lepton system, $\Delta\phi_{p_T^{\ell\ell}, p_T^{miss}} > 175°$. This retains more events in the high $p_T^{\ell\ell}$ and p_T^{ℓ} range where the sensitivity to TGC is largest. The WW signal will be selected with a purity of better than 85%. When optimized for the cross-section measurement, 110 signal events are expected to be identified at an integrated luminosity of 1 fb^{-1} [2]. In the TGC analysis the number is reduced to 75 signal events with the same amount of data.

The production of $Z\gamma \rightarrow \ell^+\ell^-\gamma$ events, within the above mentioned signal phase space, can be rather cleanly separated from all backgrounds, except those with Z bosons and real or misidentified photons in the final state. The signature of two high-p_T leptons and an isolated photon is identical. Photons outside the signal definition are however mainly from final state radiation and thus not sensitive to neutral TGCs. When rather advanced MVA techniques are applied, signal-to-background ratios of 1.7–2.0 can be reached in both the di-electron and di-muon channel with about 370 and 750 events expected in 1 fb^{-1} of data, respectively.

The process with the lowest Standard Model production cross-section is the Z boson pair production. In an ATLAS study, the $ZZ \rightarrow \ell^+\ell^-\ell^+\ell^-$ and $ZZ \rightarrow \ell^+\ell^-\nu\nu$ channels are analysed. The four-lepton final state is selected by identifying four isolated, high p_T leptons which are combined to pairs of same flavour and opposite charge. The invariant mass of each lepton pair must be compatible with the nominal Z mass within about 20 GeV and the angular separation of the leptons is required to fulfil, $\Delta R_{\ell^+\ell^-} > 0.2$. Main background is from $Zb\bar{b}$ and $t\bar{t}$ events. For this reason, hadronic isolation criteria and proper identification of the production vertex of the leptons are important to suppress events with long-lived and leptonically decaying B hadrons. Eventually, the ZZ selection efficiency reaches 24, 41 and 28% in the 4e, 4μ and $2e2\mu$ channels. This corresponds to a signal expectation of 2.6, 4.5 and 6.2 events, again in the three channels and for 1 fb^{-1} of data, with total backgrounds estimated to be below 0.3 events [2]. Relaxing the di-lepton mass requirement for one of the lepton pair and accepting also one off-shell Z boson increases the efficiency and signal expectation by about 20%, but also increases the background level to about 2 events. The ZZ final state is completed by looking for the $ZZ \rightarrow \ell^+\ell^-\nu\nu$ signal. These events are selected by asking for exactly two measured leptons of opposite charge and an E_T^{miss} value above 50 GeV. Since the ZZ pair is produced with only small p_T, the missing energy vector is pointing in opposite direction to the di-lepton momentum and is also of similar magnitude. Therefore, cuts on $|E_T^{miss} - p_T^Z| < 0.25 p_T^Z$ and $145° < \phi_{E_T^{miss}} - \phi_{p_T^{\ell\ell}} < 215°$ efficiently suppresses background from WZ events. A jet veto and a minimal $p_T^{\ell\ell}$ of 100 GeV, reduces remaining top-pair and single-Z production. Eventually, a purity of 60% is achieved and the number of expected signal events amounts to about 10 in a data sample of 1 fb^{-1}, combining the $e^+e^-\nu\nu$ and $\mu^+\mu^-\nu\nu$ channels.

The selected di-boson event samples will be used to measure the corresponding production cross-sections, and also to measure TGCs or to constrain anomalous contributions to them. When a total luminosity of 1 fb^{-1} is assumed, a signal significance of more than 5 standard deviations, typically even more than 10, are estimated in all channels, except ZZ $\rightarrow \ell^+\ell^-\nu\nu$. For the latter, about 2.5 fb^{-1} of data will be needed to establish a signal significance above 5 σ.

The main effect of anomalous TGC in di-boson production is an enhanced cross-section at high transverse momentum, p_T, of the individual gauge bosons and at high transverse masses, m_T. To test the sensitivity to TGCs, a reweighting scheme is applied. Event samples with anomalous couplings are generated with the BHO and BosoMC Monte Carlo programs in one- and two-dimensional grids of the coupling values, with step sizes in the range 10^{-4} to 10^{-3}. The differential transverse mass distribution or the p_T spectrum of one of the bosons is chosen as a reference. Each event of the main Monte Carlo sample, generated with MC@NLO, is multiplied by the ratio of the non-Standard Model and the Standard Model expectations for $d\sigma/dm_T$ (or $d\sigma/dp_T$) at a given generated m_T (p_T). The reweighted, reconstructed transverse mass distribution is then fit to the expected differential distributions determined in many sets of Monte Carlo events, each corresponding to a given integrated luminosity. In this way, the expected sensitivity is derived. Each coupling is measured in a separate fit, setting the others to their Standard Model values. Examples of such data sets and the effect of anomalous TGCs on the m_T spectrum are shown in Fig. 6.7 for the WZ and WW channels.

The Tables 6.2 and 6.3 summarise the ATLAS expectations for the determination of charged TGCs for 10 fb^{-1} of data and compares them to the LEP and Tevatron measurements. The results are based on the analysis of the m_T or p_T spectra in the different di-boson channels. Systematic uncertainties at the LHC are conservatively assumed to be in the order of 10% on the signal scale and 18% on the background normalisation. In the WZ channel, a fit to the p_T spectrum of the Z boson is

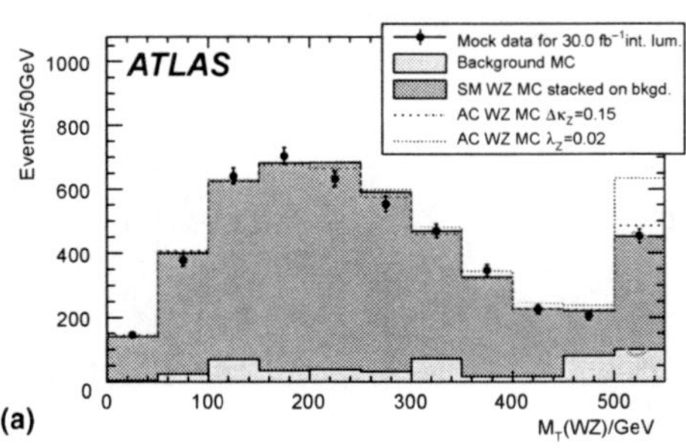

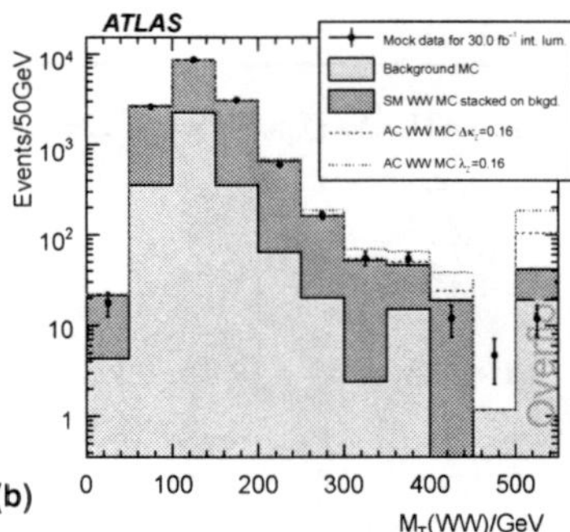

Fig. 6.7 (**a**) Simulated transverse mass distribution of the WZ system for an integrated luminosity of 30 fb^{-1} [2]. The histogram shows the Standard Model expectation, including background. The data points are derived from a randomly selected sample of Monte Carlo events that corresponds to 30 fb^{-1} of data. The dashed lines indicate the variation of the cross-section in the high transverse mass region for two different values of anomalous gauge boson couplings, $\Delta\kappa_Z = 0.15$ and $\lambda_Z = 0.02$ (**b**) The same distribution for WW events with expectations for anomalous contributions assuming $\Delta\kappa_\gamma = 0.16$ and $\lambda_Z = 0.16$, reflecting the different sensitivity to the couplings compared to the WZ channel [2]

Table 6.2 Comparison of expected sensitivities to anomalous charged TGC between W and Z bosons at the LHC for an integrated luminosity of 10 fb^{-1}. The projections are given for one experiment and the new-physics scale is chosen to be $\Lambda = 2$ TeV. The LHC expectations are compared to the combined LEP measurements [29] and the results of CDF [30] and DØ [31]. The Tevatron results are also evaluated for a Λ cut-off at 2 TeV

Process (Observable)	Limit on anomalous coupling at 95% C.L.		
	$\Delta\kappa_Z$	λ_Z	Δg_1^Z
LHC (ATLAS) expectation, $\int \mathcal{L} = 10$ fb^{-1}			
WZ (m_T^{WZ})	[−0.10, +0.22]	[−0.015, +0.013]	[−0.011, +0.034]
Wγ (p_T^{γ})	–	–	–
WW (m_T^{WW})	[−0.04, +0.07]	[−0.04, +0.04]	[−0.15, +0.31]
LEP combined, $\int \mathcal{L} = 4 \times 0.7$ fb^{-1}			
WW	–	–	[−0.051, +0.034]
DØ, all channels combined, $\int \mathcal{L} = 1$ fb^{-1}			
Wγ, WZ, WW	–	–	[−0.07, +0.16]
CDF, $\int \mathcal{L} = 1.9$ fb^{-1}			
WZ $\rightarrow \ell\nu\ell^+\ell^-$ (p_T^Z)	[−0.76, +1.18]	[−0.13, +0.14]	[−0.13, +0.23]
WZ $\rightarrow \ell\nu q\bar{q}$ (p_T^Z)	[−1.01, +1.27]	[−0.16, +0.17]	[−0.20, +0.29]

Table 6.3 Comparison of expected sensitivities to anomalous WWγ couplings at the LHC for an integrated luminosity of 10 fb^{-1}, again estimated for one experiment. The expectations are compared to the combined LEP measurements [29] and the results of CDF [31] and DØ [31]. The LHC projections and the Tevatron results are evaluated for a Λ cut-off at 2 TeV

Process (Observable)	Limit on anomalous coupling at 95% C.L.	
	$\Delta\kappa_\gamma$	λ_γ
LHC (ATLAS) expectation, $\int \mathcal{L} = 10$ fb^{-1}		
WZ (m_T^{WZ})	–	–
Wγ (p_T^{γ})	[−0.26, +0.07]	[−0.05, +0.02]
WW (m_T^{WW})	[−0.09, +0.09]	[−0.07, +0.17]
LEP combined, $\int \mathcal{L} = 4 \times 0.7$ fb^{-1}		
WW	[−0.105, +0.069]	[−0.059, +0.026]
DØ, all channels combined, $\int \mathcal{L} = 1$ fb^{-1}		
Wγ, WZ, WW	[−0.29, +0.08]	[−0.08, +0.08]

slightly less sensitive. A combined multi-dimensional analysis would improve the final result but has not yet been performed.

The LEP and Tevatron values, which are shown for comparison, are derived with the additional assumption of custodial symmetry, constraining λ_Z and κ_Z to:

$$\kappa_Z = g_1^Z - \tan^2\theta_w(\kappa_\gamma - 1) \tag{6.6}$$

$$\lambda_Z = \lambda_\gamma. \tag{6.7}$$

Setting $g_1^Z = 0$, the first condition is equivalent to $\Delta\kappa_Z = -\tan^2\theta_w\Delta\kappa_\gamma \approx -0.30\Delta\kappa_\gamma$. This enhances the sensitivity of the LEP data to these couplings since s-channel exchange of both Z and photon is involved in $e^+e^- \to$ WW production. The coupling limits measured at the Tevatron are similarly improved by the symmetry assumption since the Wγ, WZ and WW final states are combined. Furthermore, the LEP results do not take any scale dependence of the couplings into account and the couplings are thus given in the limit of zero energy scale, which is equivalent to setting the new-physics scale Λ to infinity. Applying a luminosity-averaged scale, $\sqrt{\hat{s}_{\mathrm{LEP}}}$, of about 196 GeV and a cut-off $\Lambda = 2$ TeV would increase the LEP uncertainties on charged couplings by +2% and in case of neutral couplings by +3%.

The summary Tables 6.2, 6.3, and 6.4 show that due to higher di-boson masses reached at the hadron colliders, the LEP results will be improved in future, first by the Tevatron experiments when the final expected data set of at least 2×8 fb^{-1} will have been analysed, and later by the LHC. ATLAS and CMS will then profit from an even higher centre-of-mass energy and from the fact that the prospects for

Table 6.4 Comparison of expected sensitivities to anomalous neutral TGC at the LHC for an integrated luminosity of 10 fb^{-1} analysed by one experiment using $\Lambda = 2$ TeV. The expectations are compared to the combined LEP measurements [29] and the results of CDF [30]. The Tevatron limits are evaluated for a Λ cut-off at 1.2 TeV

Process (Observable)	Limit on anomalous coupling at 95% C.L. f_4^Z	f_5^Z	f_4^γ	f_5^γ
LHC (ATLAS) expectation, $\int\mathcal{L} = 10$ fb^{-1}				
ZZ (p_T^Z)	[−0.009, +0.009]	[−0.34, +0.38]	[−0.17, +0.19]	[−0.011, +0.010]
LEP combined, $\int\mathcal{L} = 4 \times 0.7$ fb^{-1}				
ZZ	[−0.30, +0.30]	[−0.34, +0.38]	[−0.17, +0.17]	[−0.32, +0.36]
CDF, $\int\mathcal{L} = 1.9$ fb^{-1}				
ZZ	[−0.12, +0.12]	[−0.13, +0.12]	[−0.10, +0.10]	[−0.11, +0.11]
	h_3^Z	h_4^Z	h_3^γ	h_4^γ
LHC (ATLAS) expectation, $\int\mathcal{L} = 10$ fb^{-1}				
Zγ (p_T^Z)	–	–	–	–
LEP combined, $\int\mathcal{L} = 4 \times 0.7$ fb^{-1}				
Zγ	[−0.20, +0.07]	[−0.05, +0.12]	[−0.049, +0.008]	[−0.002, +0.034]
CDF, $\int\mathcal{L} = 1.1 - 2.0$ fb^{-1}				
Zγ	[−0.083, +0.083]	[−0.0047, +0.0047]	[−0.084, +0.084]	[−0.0047, +0.0047]

the integrated luminosity are in the order of several 100 fb^{-1}. Deviations from the Standard Model couplings at the per-cent level or below will be accessible. The LHC estimates presented here are expected to improve further with more involved analysis techniques – a good understanding of the detector and of the measured data provided. Therefore, the sensitivity to possible new physics at the multi-TeV range, manifesting itself in terms of modified gauge boson self-couplings, is excellent at the LHC.

6.5 Prospects for the Weak Mixing Angle

The study of electron and muon pair production at the LHC allows a measurement of the forward-backward charge asymmetry, similar to the precision measurement performed at LEP at Z peak energies. The hard scattering process, $q\overline{q} \to Z/\gamma^* \to \ell^+\ell^-$, is the inverse of the LEP reaction, when setting $\ell = e$. In the $q\overline{q}$ rest frame the differential cross-section is therefore given by Eq. (1.62) which can be simplified to:

$$\frac{d\sigma(q\overline{q} \to \ell^+\ell^-)}{d\cos\theta} \propto \left(\frac{3}{8}(1+\cos^2\theta) + A_{\rm FB}\cos\theta\right), \tag{6.8}$$

with the scattering angle, θ, of the lepton with respect to the incoming quark and with the forward-backward asymmetry, $A_{\rm FB}$, defined in the usual way:

$$A_{\rm FB} = \frac{\int_0^{+1}\frac{d\sigma}{d\cos\theta}d\cos\theta - \int_{-1}^{0}\frac{d\sigma}{d\cos\theta}d\cos\theta}{\int_{-1}^{+1}\frac{d\sigma}{d\cos\theta}d\cos\theta} = \frac{\sigma_F - \sigma_B}{\sigma_F + \sigma_B} = \frac{N_F - N_B}{N_F + N_B}. \tag{6.9}$$

From a measurement of $A_{\rm FB}$, the weak mixing angle, $\sin^2\theta^{\ell}_{\rm eff}$, can be extracted. This is particularly interesting because the LEP and SLC data show some difference between the mixing angles measured in leptonic and hadronic final states (see Chap. 4). A similar precision as LEP/SLC of $\delta\sin^2\theta^{\ell}_{\rm eff} \approx 2\times 10^{-4}$ or better is therefore the ultimate goal. Currently, the Tevatron experiments reach accuracies of $\delta\sin^2\theta^{\ell}_{\rm eff} = 1.9\times 10^{-3}$ [32] using about 1 fb^{-1} of data, mainly dominated by statistical uncertainties and fully compatible with the LEP results.

The $A_{\rm FB}$ measurement also gives a handle to constrain new physics beyond the Standard Model, like additional heavy vector bosons. Their existence may alter $A_{\rm FB}$ in regions of invariant lepton pair masses which are below the mass of the new vector boson due to interference effects [33].

At the LHC, the scattering angle is not identical to the angle of the lepton with respect to the beam axis because the $q\overline{q}$ system is not produced at rest and carries transverse and longitudinal momentum with respect to the beam directions. The transverse momentum leads to an ambiguity when calculating the quark and anti-quark momenta from the kinematics of the $\ell^+\ell^-$ system. The scattering angle is therefore determined in the Collins–Soper frame [34] which reduces this ambiguity.

The particle four-momenta are Lorentz boosted into the $\ell^+\ell^-$ rest frame and the polar axis is chosen to be the bisection of the two proton momentum vectors. The scattering angle, θ^*, with respect to the polar axis is then given by [35]

$$\cos\theta^* = \frac{P_L}{|P_L|}\frac{2}{P^2\sqrt{P^2+P_T^2}}\left(P_1^+P_2^- - P_1^-P_2^+\right), \tag{6.10}$$

with the four-momentum, P, and the transverse and longitudinal momenta, P_T and P_L, of the lepton pair. The quantities $P_i^{\pm} = \frac{1}{\sqrt{2}}(E_i \pm p_{L,i})$ $(i = 1, 2)$ are calculated from the energies, E_i, and longitudinal momenta, $p_{L,i}$, of the lepton and the anti-lepton. The $\cos\theta^*$ distribution, expected for the ATLAS experiment, is shown in Fig. 6.8.

The sign of $\cos\theta^*$, respectively the quark direction, can only be inferred indirectly in pp collisions. This is different from the situation at $p\bar{p}$ colliders where the quark can be identified with one of the valence-quarks of the proton, and the quark and proton beam directions are thus the same. At the LHC, the quark is assumed similarly to be a valence-quark of the proton and sea contributions are small if the $\ell^+\ell^-$ pair is not produced centrally. The anti-quark, however, is necessarily from the sea-quarks inside the proton. It carries therefore a much lower momentum fraction than the quark in the $q\overline{q}$ reaction. Under these assumptions, the quark direction is preferably oriented along the direction of the boosted Z/γ^* system [36]. In particular, the sign of the longitudinal component of the lepton-pair momentum indicates the hemisphere of the quark direction, which is taken into account by the additional factor $P_L/|P_L|$ in Eq. (6.10).

For a precise measurement of the charge asymmetry, a large angular coverage of the detector is necessary. In an ATLAS study [2], electron pairs identified in the central ($|\eta| < 2.5$) and forward ($2.5 < |\eta| < 4.9$) calorimeters are used. Muons

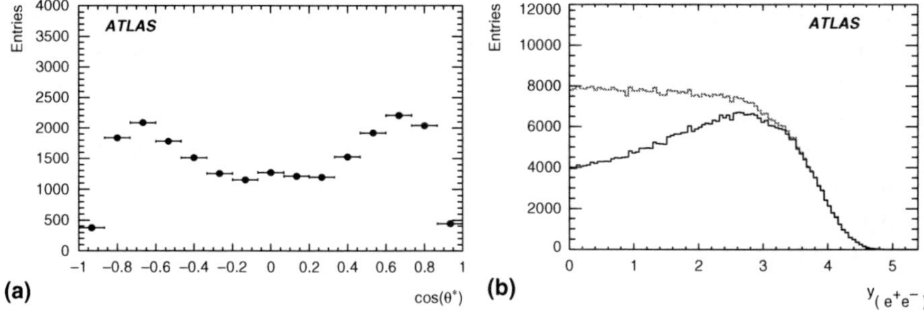

Fig. 6.8 (**a**) Distribution of $\cos\theta^*$ for reconstructed electron pair events in the Z pole region [2]. The monte carlo events are generated with the PYTHIA [37] program and passed through the ATLAS detector simulation. (**b**) Simulated rapidity distribution of e^+e^- pairs [2]. The upper line shows the distribution for all events, the lower line only events with correctly reconstructed quark direction

are not being studied since the corresponding detector acceptance is restricted to $|\eta| < 2.7$ only.

The electrons are required to pass p_T thresholds of about 20 GeV. To identify the charge and the direction of the e^-, respectively e^+, at least one of the electrons must have a corresponding track measured in the inner tracking detectors. Shower shape criteria optimised for central and forward electrons improve the purity of the selected event sample. The mass of the reconstructed e^+e^- system is restricted to the Z mass region, $|\sqrt{P^2} - M_Z| < 12$ GeV. Small missing transverse energy reduces background with neutrinos in the final state. The rapidity of the e^+e^- pair should not be central, $|y_{e^+e^-}| > 1$, to suppress $q\bar{q}$ reactions of only sea-quarks and to increase the probability of a correct identification of the quark direction, as illustrated in Fig. 6.8. Main backgrounds are QCD di-jet, W+jet, and $t\bar{t}$ events. The background events contribute at most with 0.3% to the final sample, mainly in the configuration with one central and one forward electron.

If the e^+e^- mass interval is restricted to the Z pole region, the dependence of the forward-backward asymmetry on $\sin^2\theta^\ell_{\text{eff}}$ can be well approximated by a linear function [38]:

$$A_{\text{FB}} = b\left(a - \sin^2\theta^\ell_{\text{eff}}\right), \tag{6.11}$$

with parameters a and b which include radiative corrections and depend in particular on the parton distribution functions. At LHC energies of 14 TeV and using the PYTHIA Monte Carlo interfaced to PDFs provided by the MRST group [39] one obtains $a = 0.23 \pm 0.03$ and $b = 1.83 \pm 0.26$ [2]. Using the above formula, the weak mixing angle can be directly extracted from A_{FB}.

The measurement of the charge asymmetry may be altered by systematic effects which are studied as well. The uncertainty on the charge determination can be derived from the number of electron pairs in the Z peak region which are measured with charges of equal sign. The expected charge misidentification rate is small, up to a maximum of 0.015 at $|\eta| = 2.5$, decreasing with smaller pseudo-rapidity values. It results in an uncertainty on A_{FB} of about 3×10^{-5}. Experimental uncertainties on energy scale, energy resolution, and reconstruction efficiency contribute with 4×10^{-5} to the systematics. Since the background rates are small in the Z peak region, the corresponding uncertainty is below 10^{-5} on A_{FB}. The total systematics is therefore on the level of 5×10^{-5}, to be compared with the expected statistical precision of $\delta A_{\text{FB}} \approx 3 \times 10^{-4}$ calculated for a total luminosity of 100 fb^{-1}. This corresponds to 2–3 years of high-luminosity running of LHC.

Applying relation (6.11) the weak mixing angle is derived from the A_{FB} measurement. Due to the factor $b \approx 1.8$ the uncertainties on $\sin^2\theta^\ell_{\text{eff}}$ are practically halved with respect to those of A_{FB}. The expected statistical precision is therefore in the order of 1.5×10^{-4}. The dominant systematic effect is however not one of those discussed above, but the uncertainties on the PDFs which are used to relate A_{FB} to the underlying theory. The systematics is estimated by running the event simulation with different MRST data sets representing the uncertainties assigned to the PDFs. This yields an uncertainty of about 2×10^{-4} on $\sin^2\theta^\ell_{\text{eff}}$. The systematic uncertainty on the parton distributions inside the proton is also the dominant systematic effect in the corresponding measurement of $\sin^2\theta^{\ell}_{\text{eff}}$ at the Tevatron [32, 40].

Improving the knowledge on the event modelling is therefore an important ingredient to control and possibly reduce the systematic effects on the weak mixing angle. In addition, higher order QCD corrections at NLO and beyond need to be applied to the signal prediction, which can be obtained, e.g., by event reweighting [32]. The LHC experiments will be able to perform a measurement of $\sin^2\theta^{\ell}_{\text{eff}}$ competitive with the LEP/SLC results if experimental and theoretical uncertainties can be mastered. The large data sets will help to study these in great detail by using different control and physics samples, for example $W \rightarrow e\nu$ events which may be used to narrow down the shape of the PDFs (see Chap. 1).

6.6 Prospects for Electroweak Measurements at the LHC

In summary, assuming a reasonable performance of the ATLAS and CMS detectors, many electroweak parameters will be measured more precisely. The experimental challenges in the top and W mass measurements are similar to those experienced at the Tevatron. High luminosity runs will help to reduce systematics by either studying the different sources better or by constraining the data sets to phase space regions which are less affected by those. This will also lead to a new precision measurement of the weak mixing angle. Due to the higher centre-of-mass range, improved determinations of the triple-gauge boson couplings will be possible. The LHC data will therefore push the experimental tests of the Standard Model predictions to even higher precision.

On the other hand, new physics may be discovered in this indirect way. Due to the LHC potential to observe new effects directly, like super-symmetric particles or heavy vector bosons, the precision measurements may be linked and compared with the new theoretical models which may manifest in proton–proton collisions at 14 TeV. But even if the Standard Model does not need to be extended in the energy range reached at the LHC, the Higgs boson should be detected by the LHC experiments, as the last missing building block of the Standard Model and manifestation of electroweak symmetry breaking. The indirect and direct Higgs mass determination could then be compared again as a further proof or disproof of the consistency of the theoretical model. The search for the Standard Model Higgs boson at the LHC is thus very important and subject of the following chapter.

References

1. K. Melnikov and F. Petriello, Phys. Rev. Lett. **96** (2006) 231803.
2. The ATLAS Collaboration, G. Aad et al., *Expected Performance of the ATLAS Experiment Detector, Trigger, Physics*, CERN-OPEN-2008-020, arXiv:0901.0512.
3. The ATLAS Collaboration, *ATLAS Forward Detectors for Measurement of Elastic Scattering and Luminosity*, ATLAS TDR 018, CERN/LHCC 2008-04.
4. M. L. Mangano, M. Moretti, F. Piccinini, R. Pittau and A. Polosa, JHEP **0307** (2003) 001.
5. J. M. Campbell, J. W. Huston and W. J. Stirling, Rept. Prog. Phys. **70** (2007) 89.

6. N. Besson, M. Boonekamp, E. Klinkby, T. Petersen, S. Mehlhase, for the ATLAS Collaboration, Eur. Phys. J. **C 57** (2008) 627; arXiv:0805.2093.
7. M. Awramik, M. Czakon, A. Freitas, G. Weiglein, Phys. Rev. **D 69** (2004) 053006; arXiv:hep-ph/0311148
8. The ALEPH Collaboration, the DELPHI Collaboration, the L3 Collaboration, the OPAL Collaboration, the SLD Collaboration, the LEP Electroweak Working Group, the SLD electroweak, heavy flavour groups, Phys. Rept. **427** (2006) 257; hep-ex/0509008v3.
9. The CMS Collaboration, G.L. Bayatian et al., CMS Physics Technical Design Report, Vol. II: Physics Performance, CERN/LHCC 2006-021.
10. J. Pumplin, D. R. Stump, J. Huston, H. L. Lai, P. Nadolsky and W. K. Tung, JHEP **0207** (2002) 012, hep-ph/0201195; J. Pumplin et al., *Parton Distributions and the Strong Coupling Strength: CTEQ6AB PDFs*, hep-ph/0512167.
11. R. Bonciani et al., Nucl. Phys. **B 529** (1998).
12. N. Kidonakis et al., Phys. Rev. **D 68** (2003) 114014.
13. U. Langenfeld, S. Moch, P. Uwer, *New Results for $t\bar{t}$ Production at Hadron Colliders*, Preprint DESY-09-104, 2009, arXiv:0907.2527.
14. P. M. Nadolsky et al., Phys. Rev. **D 78** (2008) 013004; W. Beenakker et al., Phys. Rev. **D 40** (1989) 54.
15. M. Cacciari et al., JHEP **09** (2008) 127.
16. S. Moch and P. Uwer, Phys. Rev. **D 78**, (2008) 034003; U. Langenfeld, S. Moch and P. Uwer, *Measuring the running top-quark mass*, Preprint DESY 09-097, 2009, arXiv:0906.5273.
17. N. Kidonakis and R. Vogt, Phys. Rev. **D 78** (2008) 074005.
18. DØ Collaboration, V. Abazov et al., *Combination and Interpretations of ttbar Cross Section Measurements with the DØ Detector*, Preprint FERMILAB-PUB-09-092-E, 2009, arXiv:0903.5525.
19. The CDF Collaboration, *Combination of CDF Top Quark Pair Production Cross Section Measurements with* 2.8 fb^{-1}, CDF Note 9448, 2008.
20. Tevatron Electroweak Working Group, for the CDF Collaboration, the DØ Collaboration, *Combination of CDF and D0 Results on the Mass of the Top Quark*, arXiv:0903.2503v1.
21. J. Campbell, R. K. Ellis and F. Tramontano, Phys. Rev. **D 70** (2004).
22. S. Frixione and B. R. Webber, JHEP **0206** (2002) 029; S. Frixione et al., JHEP **0308** (2003) 007.
23. B. P. Roe, H.-J. Yang, J. Zhu, Y. Liu, I. Stancu and G. McGregor, Nucl. Inst. Meth. **A 543** (2005) 577; physics/0408124.
24. The CDF Collaboration, T. Aaltonen et al., *First Observation of Electroweak Single Top Quark Production*, Preprint FERMILAB-PUB-09-059-E, 2009, arXiv:0903.0885; J. Swain and L. Taylor, Phys. Rev. **D 58** (1998) 093006.
25. DØ Collaboration, V. M. Abazov et al., *Observation of Single Top-Quark Production*, arXiv:0903.0850.
26. S. Frixione and B. R.Webber, JHEP **0206** (2002) 029; S. Frixione, P. Nason and B. R.Webber, JHEP **0308** (2003) 007.
27. M. Dobbs and M. Lefebvre, *Unweighted Event Generation in Hadronic WZ Production at the First Order 30 in QCD*, ATLAS Note ATL-PHYS-2000-028, 2000.
28. U. Baur, T. Han and J. Ohnemus, Phys. Rev., **D 50** (1994) 1917; U. Baur, T. Han and J. Ohnemus, Phys. Rev. **D 51** (1995) 3381; U. Baur, T. Han and J. Ohnemus, Phys. Rev. **D 53** (1996) 1098; U. Baur, T. Han and J. Ohnemus, Phys. Rev., **D 57** (1998) 2823.
29. The LEP Collaborations ALEPH, DELPHI, L3, OPAL, and the LEP Electroweak Working Group, *A Combination of Preliminary Electroweak Measurements and Constraints on the Standard Model*, CERN-PH-EP/2006-042; hep-ex/0612034.
30. http://www-cdf.fnal.gov/physics/ewk/, public notes are regularly appearing.
31. The DØ Collaboration, V. Abazov et al., Phys. Rev. Lett. **100** (2008) 131801, arXiv.org:0712.0599; The DØ Collaboration, V. Abazov et al., Phys. Lett. **B 653** (2007) 378, arXiv.org:0705.1550; The DØ Collaboration, V. Abazov et al., Phys. Rev. **D 74** (2006)

057101, hep-ex/0608011; The DØ Collaboration, V. Abazov et al., *Measurement of the WW Production Cross Section with Dilepton Final States in p-pbar Collisions at sqrt(s)=1.96 TeV and Limits on Anomalous Trilinear Gauge Couplings*, arXiv:0904.0673v1; The DØ Collaboration, V. Abazov et al., Phys. Rev. **D 76** (2007) 111104, arXiv.org:0709.2917.
32. DØ Collaboration, V.M. Abazov et al., Phys. Rev. Lett. **101** (2008) 191801, arXiv:0804.3220.
33. J. L. Rosner, Phys. Rev. **D 54** (1996) 1078, hep-ph/9512299.
34. J. C. Collins and D. E. Soper, Phys. Rev. **D 16** (1977) 2219.
35. U. Bau, S. Keller, W. K. Sakumoto, Phys. Rev. **D 57** (1998) 199, hep-ph/9707301.
36. M. Dittmar, Phys. Rev. **D 55** (1997) 161.
37. T. Sjöstrand, P. Edén, C. Friberg, L. Lönnblad, G. Miu, S. Mrenna and E. Norrbin, Comp. Phys. Comm. **135** (2001) 238; T. Sjöstrand, *Recent Progress in PYTHIA*, Preprint, LU-TP-99-42, hep-ph/0001032.
38. J. L. Rosner, Phys. Lett. **B 221** (1989) 85.
39. MRST Collaboration, Phys. Lett. **B 652** (2007) 292; A. D. Martin, W. J. Stirling, R. S. Thorne and G. Watt, *Parton Distributions for the LHC*, Preprint IPPP/08/95,arXiv:0901.0002.
40. CDF Collaboration, D. Acosta et al., Phys. Rev. **D71** (2005) 052002, hep-ex/0411059.

Chapter 7
Higgs Physics at the LHC

The search for Higgs bosons is one of the main goals of the LHC physics program. In the Standard Model there is only one neutral Higgs boson predicted. Supersymmetric extensions of the Standard Model require at least two Higgs doublets which manifest as three neutral and two charged Higgs bosons. In the following chapter, mainly the search for the Standard Model Higgs boson and the possibilities to measure its properties at LHC are described. The latter subject is however also relevant for the possible discovery of more than one neutral Higgs particle, since these are expected to differ in mass, coupling structure and CP properties.

7.1 Standard Model Higgs Searches with ATLAS and CMS

The search for SM Higgs bosons [1] is one of the primary goals of the LHC experiments. Present direct search results for the SM Higgs exclude masses, M_{H}, below 114.4 GeV [2] and M_{H} values in the range 160–170 GeV at 95% C.L. [3]. Global Standard Model analyses of electroweak data prefer Higgs mass values below 186 GeV [4], as discussed in Chap. 4. In the given mass range, the SM Higgs is mainly produced through gluon–gluon fusion, $gg \to H$, where subsequent $H \to \gamma\gamma$, $H \to \mathrm{ZZ} \to 4\ell\,(\ell = \mathrm{e}, \mu)$ and $H \to \mathrm{WW} \to \ell\nu\ell\nu, \mathrm{qq}\ell\nu$ decays are analysed. Figure 7.1 compiles the SM Higgs branching fractions and cross-sections for $\sqrt{s} = 14$ TeV. They show that the two-photon decay can only be explored up to Higgs masses of about 120 GeV, while di-boson decays start to have a significant rate below the di-boson mass threshold. The t-channel vector-boson-fusion (VBF) process has a much lower production rate. However, it has experimentally useful signatures due to the event kinematics with two forward quark jets, suppressed central jet production and a central Higgs decay. In VBF, Higgs decays to WW and $\tau^+\tau^-$ are studied. In the very low Higgs mass region, $H \to \mathrm{b\overline{b}}$ dominates, but is only detectable in Higgs-Strahlung from a $\mathrm{t\bar{t}}$ pair, leading to a $\mathrm{t\bar{t}}H \to \mathrm{t\bar{t}b\overline{b}}$ final state.

The search for $gg \to H \to \gamma\gamma$ is very challenging. The di-photon $\mathrm{q\overline{q}}, \mathrm{q}g \to \gamma\gamma + X$ and $gg \to \gamma\gamma$ backgrounds are irreducible and dominate the spectrum together with the QCD two-jet background with misidentified jets. Only a good mass resolution helps to identify the Higgs signal, as shown in Fig. 7.2 for simulated

A. Straessner, *Electroweak Physics at LEP and LHC*, STMP 235, 189–207,
DOI 10.1007/978-3-642-05169-2_7, © Springer-Verlag Berlin Heidelberg 2010

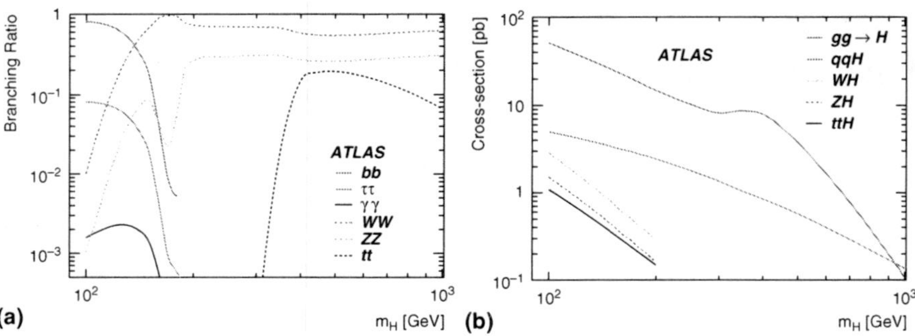

Fig. 7.1 (**a**) SM Higgs branching fractions as a function of the Higgs mass, M_{H} [5]. (**b**) Cross-sections of different SM Higgs production modes at the LHC using NLO calculations [5].

ATLAS data. Since there is a significant amount of material in the ATLAS detectors in front of the LAr calorimeter, reconstruction of photon conversions is necessary. Some 57% of the selected events have at least one conversion at a radius smaller than 80 cm from the interaction point. The single and double track conversion reconstruction recovers those photons with efficiencies between 40 and 90% depending on the conversion radius.

The photons are reconstructed in the electromagnetic calorimeter as clusters combining energy depositions in LAr cells in given $\Delta\eta \times \Delta\phi$ areas. In the LAr barrel region, $|\eta| < 1.37$, the cluster sizes cover 3×7 cells of the middle layer of the calorimeter in case the photons are tagged as converted, while unconverted photons are identified as 3×5 clusters. In the LAr endcaps, $1.52 < |\eta| < 2.37$, a single cluster size of 5×5 is chosen. Shower shape and and measurement of energy leakage in the first compartment of the Tile Calorimeter, as well as π^0 rejection in the LAr strips are used to reject hadronic background. The detection efficiency of photons with $p_T > 25$ GeV is about 83% when low-luminosity pile-up at 10^{33} cm^{-2}s^{-1} is included. Furthermore, isolation from nearby tracks improves the rejection of hadronic jets to a factor of about 8,000 [5].

Also the primary vertex reconstruction is important since it influences the calculated Higgs mass. Due to the high granularity of the multi-layer calorimeter, the cluster bary-centres provide information about the direction of impact of the photons. Combined with a primary vertex constraint of the tracking system, the $H \to \gamma\gamma$ decay vertex can be measured with a resolution of about 0.1 μm.

The mass resolution obtained after applying the full reconstruction of the $H \to \gamma\gamma$ decay is better than 1.5% which is illustrated in Fig. 7.2. The effect of the photon triggers is already included. These are efficient to more than 94% for di-photon p_T thresholds of 17 GeV. Here, the offline event selection is taken as a reference which requires that the photons are detected within the fiducial volume of the LAr calorimeter excluding the low-efficiency gap between barrel and endcap, and that the p_T of the two photons is larger than 25 and 40 GeV, respectively.

The expected performance of the inclusive Higgs boson detection is evaluated within a mass window of $\pm 1\ \sigma$ around the peak value, where σ is the mass

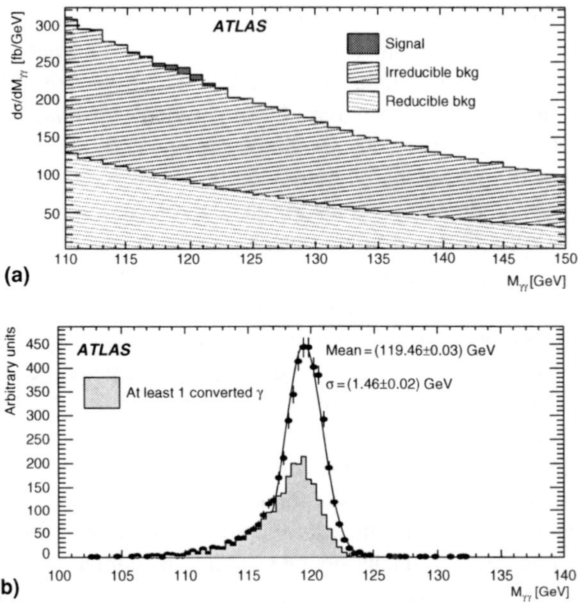

Fig. 7.2 (**a**) Di-photon invariant mass spectrum after the application of cuts of the inclusive analysis, without additional requirements on jets, E_T^{miss} or leptons [5]. (**b**) Invariant mass distribution for photon pairs from $H \to \gamma\gamma$ decays with $M_{\text{H}} = 120$ GeV after trigger and identification cuts [5]. Events with at least one converted photon are displayed as the shaded histogram

resolution. For a Higgs mass of 120 GeV, the accepted cross-section for the signal is 25.4 fb, combining gluon-fusion production, vector-boson fusion, $\text{t}\bar{\text{t}}H$ and WH/ZH associated production. The background amounts to 950 fb, dominated by the irreducible QCD di-photon processes (53%) and by γ+jet production (33%). This shows that a very good performance and resolution of the LAr calorimeter will be important not to miss the $H \to \gamma\gamma$ signal.

The signal-to-background ratio can be further improved to about 1/10 and 1/2 by requiring one or two additional jets, respectively, because the leading jet in $gg \to H + \text{jet}$ and VBF production tends to be harder than for the di-photon background process. This reduces the accepted signal by factors 5 and 25, though. Other search strategies require E_T^{miss} or E_T^{miss} + lepton signatures, which enhances the associated production WH, ZH, $\text{t}\bar{\text{t}}H$ with respect to the background. The signal cross-sections are however below 0.01 fb, which means that these channels become significant only if a larger amount of data, several hundred fb^{-1}, will be collected.

A similar performance is also achieved by the CMS experiment. The details of the measured photon shower shape in the CMS crystal calorimeter in different η ranges is used to divide the data into performance categories. By optimising the analysis using cut-based and neural network criteria for photon selection and hadronic isolation, signal-to-background ratios close to 1/1 can be achieved in some of the event categories [6].

Another benchmark channel, where well performing calorimetry is essential, is the $H \to ZZ \to 4\ell$ channel. Here, high p_T electrons and muons are triggered on, and four leptons compatible with two Z decays, possibly off-shell, are selected. Electrons are reconstructed as clusters in the electromagnetic calorimeters matched to tracks in the inner detector. Shower shape and track quality criteria are applied. Isolation from hadronic activities is important to reject $Zb\bar{b} \to 4\ell + X$ and $t\bar{t} \to 4\ell + X$ production. Muons are detected in the muon spectrometer and a combined reconstruction with inner detector tracks is performed. Also here, isolation criteria rejecting muons with nearby tracks and clusters are used to reduce background. Both electrons and muons may be produced by B meson decays with vertices displaced from the primary vertex. To enhance the Higgs signal and reduce background with b quark content, cuts on the impact parameter are applied to muons and electrons. Eventually, lepton pairs of opposite charge, e^+e^- and $\mu^+\mu^-$, are are combined and the $H \to ZZ \to 4\ell$ decay is fully reconstructed. The trigger efficiency for the four-lepton final states are typically above 98% compared to the offline event selection since either single-lepton triggers with p_T thresholds of about 20 GeV and isolation criteria or two-lepton signature with lower thresholds of 10–15 GeV with and without isolation are selecting the signal events.

The continuum ZZ production represents the largest background in this channel. In the final step of the analysis, the invariant mass calculated from the four-momenta of the final state leptons is the main observable used to reject this background. The mass resolution is found to be below the 2% level in ATLAS for Higgs masses smaller than 200 GeV. Figure 7.3 shows the reconstructed mass peak in a data sample of 30 fb^{-1}. At lower luminosities of around 1 fb^{-1}, additional QCD jet, weak boson, and photon production backgrounds become important because dedicated studies are needed to control the systematics uncertainties related to those background sources. The reduction and control of the ZZ continuum systematics is however primordial [7] and will require a good understanding of the detector performance and resolution with first data. The signal significances expected for the ATLAS analyses [5] in the $H \to ZZ$ channels are summarised in Fig. 7.3. The highest significances are obtained in the Higgs mass regions between 130 GeV and 160 GeV as well as between 200 and 500 GeV. The gap is mainly due to the rising ZZ continuum cross-section at about twice the Z boson mass.

In the mass range between 150 and 180 GeV the Higgs boson decays nearly exclusively to W-boson pairs. It is studied for both gluon-fusion, $H \to WW \to e\nu\mu\nu$ without additional jets, and vector-boson fusion production, $qqH \to qqWW \to qqe\nu\mu\nu$ and $qqH \to qqWW \to qqqq\ell\nu$ with additional jets in the forward region. The former is largely dominating but requires the rejection of final states with two leptons and missing energy, like W-pair, W+jets, $t\bar{t}$, $Z \to \tau^+\tau^-$ and heavy quark pair-production. For the latter, the same background sources are important, but the identification of two hadronic jets with a pseudo-rapidity gap of $|\Delta\eta_{jj}| > 2.5$ and a central jet veto further reduces those. This is due to the special VBF kinematics, as discussed in Chap. 1. To control the $t\bar{t}$ background a veto on b-jets is furthermore applied under the condition that the performance of the b-tagging algorithms is well understood.

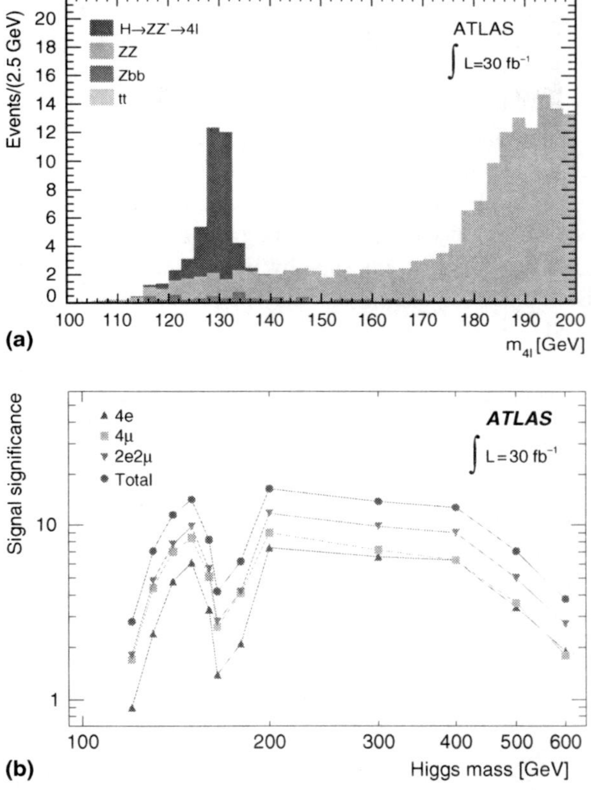

Fig. 7.3 (**a**) Reconstructed 4-lepton mass for SM Higgs signal and background, which is mainly from ZZ continuum production [5]. (**b**) Expected signal significances in the $H \rightarrow$ ZZ channel, computed using Poisson statistics, for each of the three final states, and their combination [5]

Figure 7.4 summarises the ATLAS expectations [5] for the SM Higgs discovery potential with 10 fb^{-1} and the possible exclusion for 2 fb^{-1} of data. The most sensitive mass range is around 160 GeV. However, this mass range is currently already well tested by CDF and DØ [3]. An exclusion or first evidence for a SM Higgs signal will most probably be in experimental reach of the Tevatron experiments before ATLAS or CMS have collected enough data to become sensitive in this mass range.

In the Higgs mass interval 114.4 GeV $< M_{\text{H}} <$ 120 GeV, a discovery is most challenging and more luminosity may be needed. Higgs decays to $\text{b}\bar{\text{b}}$ dominate but are practically impossible to detect with sufficient significance in direct production due to overwhelming QCD background. Associated production of Higgs bosons in the $\text{t}\bar{\text{t}}$H, ZH and WH channels provides additional signatures to identify the signal process. Higgs production with top pairs is studied by selecting events with at least 4 b-jets, two from the Higgs decay and two from the t→bW decays, and identified hadronic or leptonic W decays. The main background source is QCD

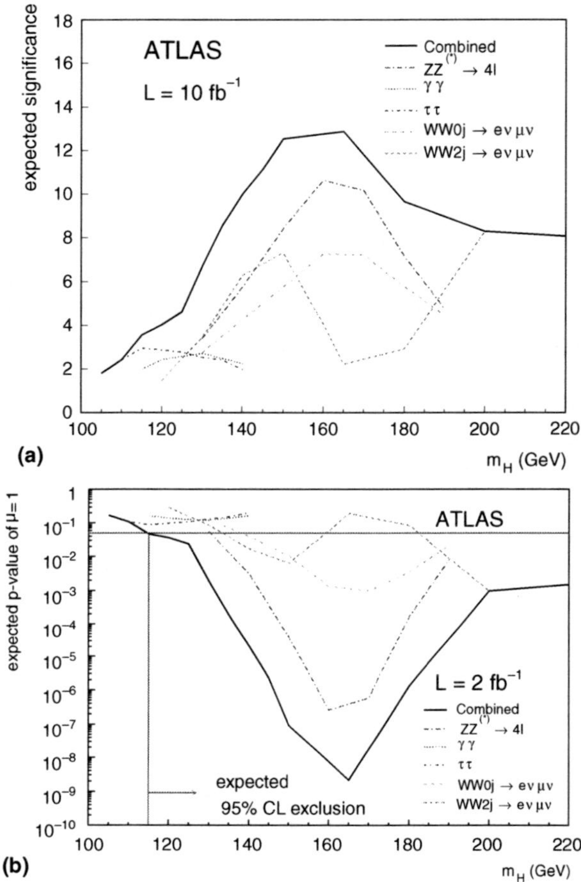

Fig. 7.4 (**a**) The median discovery significance for the various Higgs decay channels and their combination for $\mathcal{L} = 10\ \text{fb}^{-1}$ in the low M_{H} region [5]. Requiring significances above 5 translates into a possible discovery of the SM Higgs for $M_{\text{H}} > 126$ GeV. (**b**) The median p-value obtained in a profile likelihood analysis [5] for excluding a Standard Model Higgs boson ($\mu = \sigma_{\text{excluded}}/\sigma_{\text{SM}} = 1$) for the various channels, as well as the combination for the lower mass range. Assuming $\mathcal{L} = 2\ \text{fb}^{-1}$ and asking for $p > 0.95$, it will be possible to exclude the existence of the Standard Model Higgs boson in the mass range 115 GeV $< M_{\text{H}} <$ 480 GeV

and electroweak $\text{t}\bar{\text{t}}\text{b}\bar{\text{b}}$ production, but also $\text{t}\bar{\text{t}}$ events with additional hadronic jets not from B decays. They remain irreducible because requirements on mass windows for the reconstructed W and top decays only reduce other backgrounds efficiently. Eventually, signal significances in the order of $0.5 - 1\sigma$ are obtained with 30 fb^{-1} of data [6, 5], depending on the systematic uncertainties assumed. These are mainly driven by b-tag performance and b-jet misidentification rates, as well as energy scales for leptons, jets and missing transverse energy. Estimations of total background uncertainties, including predictions of the corresponding cross-sections, are in the order of 30%.

Promising results [8] are obtained in the WH→ $\ell\nu$ + $\mathrm{b\bar{b}}$ and ZH→ $\ell^+\ell^-$ + $\mathrm{b\bar{b}}$, $\nu\bar{\nu}$+$\mathrm{b\bar{b}}$ channels using a dedicated identification method of the $H \rightarrow \mathrm{b\bar{b}}$ system. If the Higgs boson is produced with large transverse momentum, $p_T > 200$ GeV, and back-to-back to the gauge boson, background from $\mathrm{t\bar{t}}$ and QCD events is suppressed. The boosted $H \rightarrow \mathrm{b\bar{b}}$ decay is however difficult to detect since the b-jets are merged into a single jet if classical jet-cone algorithms are applied. The jet substructure can however be further resolved, for example, by applying the Cambridge/Aachen [9] clustering algorithm. It is based on angular distances in $\Delta R = \sqrt{(\Delta\eta)^2 + (\Delta\phi)^2}$ between two clusters and iteratively merges objects close in angular space until all ΔR are larger than a given parameter R_{cut}. To resolve the jet substructure the last steps of the jet clustering are undone, until the jet that was formed last splits into two jets of nearly equal mass. The last condition should suppress soft gluon radiation. If both jets were tagged as a b-jet the event is considered to be a Higgs candidate. After having identified the $\mathrm{b\bar{b}}$ system, the Cambridge/Aachen algorithm is repeated with a finer resolution parameter to resolve also a possible third jet from final state QCD radiation. The accompanying gauge boson decays are identified according to their two-lepton, one lepton and E_T^{miss}, or large E_T^{miss} signatures. Events with other jets or leptons with large p_T in the central detector region, $|\eta| < 2.5 - 3$, are rejected. Eventually, one obtains a sufficient Higgs mass resolution and background suppression with signal significances in the order of 4–6 standard deviations for Higgs masses of $M_{\mathrm{H}} \approx 115$ GeV.

Figure 7.5 shows the reconstructed $H \rightarrow \mathrm{b\bar{b}}(g)$ mass spectrum for all WH and ZH channels combined. B-tag efficiencies of 60% and a mistag probability of 2% are furthermore assumed. The dependence of the final significance on these parameters, as well as on the minimal p_T of the Higgs is also indicated in Fig. 7.5. The significance decreases with increasing Higgs masses, increasing Higgs p_T and

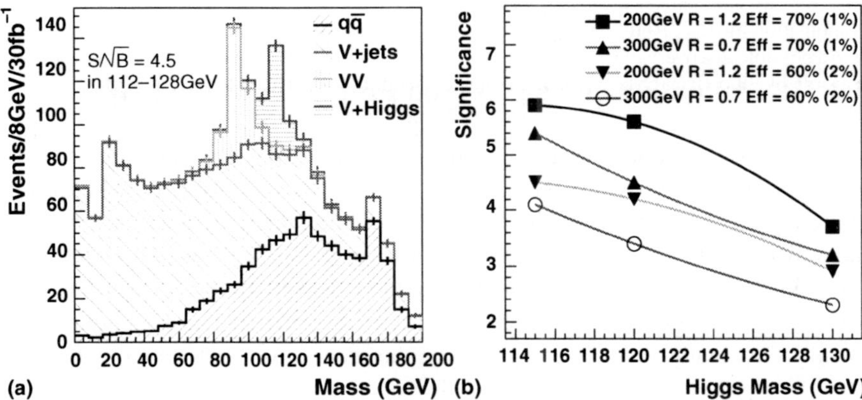

Fig. 7.5 (**a**) Reconstructed $H \rightarrow \mathrm{b\bar{b}}(g)$ mass spectrum in the combined WH and ZH search channels, expected for an integrated luminosity of 30 fb^{-1} and a LHC centre-of-mass energy of 14 TeV [8]. (**b**) Combined WH and ZH signal significance as a function of Higgs mass, of the minimal p_T of the $H \rightarrow \mathrm{b\bar{b}}(g)$ system, and of the b-tagging efficiency and mistag rate [8]

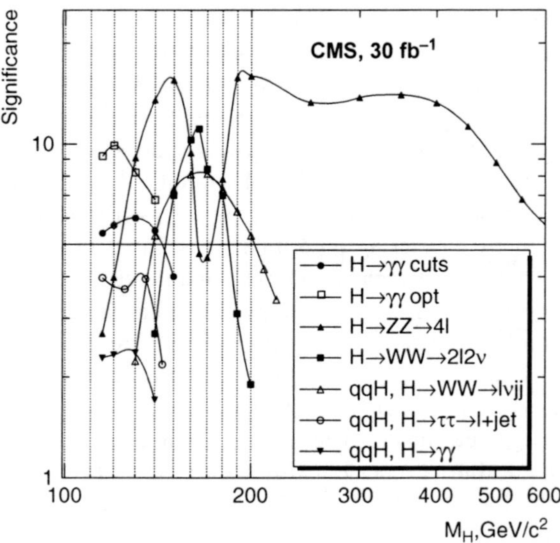

Fig. 7.6 Signal significance as a function of the Higgs boson mass for an integrated luminosity of 30 fb^{-1} for different Higgs boson production and decay channels as expected for the CMS experiment [6]

increasing b-mistag rate. The performance is however better than expected for the $t\bar{t}H$ channel and very promising in the low Higgs mass region, which is preferred by fits to electroweak data.

In general, with the full data set of ATLAS, the SM Higgs can not escape detection. The performance of CMS is quite similar, as shown in Fig. 7.6 and both LHC experiments will cross-check each others findings. The LHC data will eventually give the definitive verdict about the SM Higgs boson.

7.2 Determination of Higgs Boson Properties at the LHC

Once a Higgs boson has been discovered a determination of the particle properties will be needed to verify their consistency with the Standard Model predictions. By analysing the mass spectrum, the cross-sections and branching fractions, the Higgs mass, width and the Higgs couplings can be measured. The detailed production and decay kinematics will give a handle to determine the spin of the discovered particle and its CP properties.

7.2.1 Measurement of Mass, Width and Couplings

The first parameter which will be know at the moment of the discovery of the Higgs boson is its mass. However, the corresponding precision varies with Higgs mass and

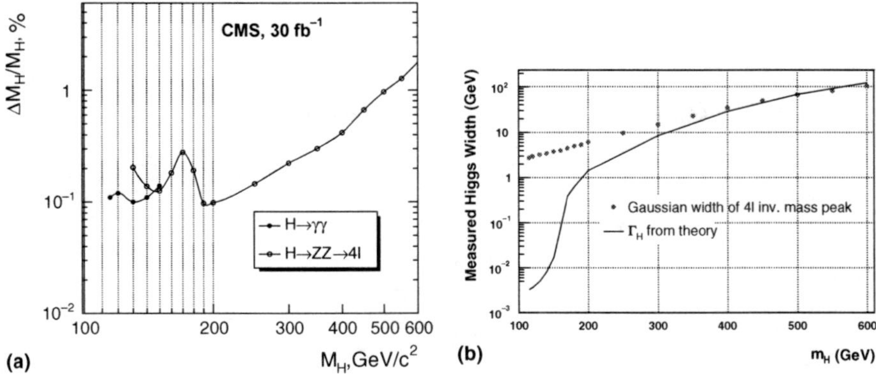

Fig. 7.7 (**a**) The determination of the Higgs mass is best in the $H \to \gamma\gamma$ and $H \to$ ZZ decay channels since the full final state can be reconstructed and the decay spectrum is measured with good resolution. (**b**) The measurement of the Higgs width is also derived from the reconstructed Higgs decay spectrum. Because of the small physical Higgs width the experimental data will only be able to measure Γ_H for Higgs masses above ≈ 200 GeV. The corresponding simulated performances are shown for the CMS experiment [6]

accessible decay channels. As can be seen in Fig. 7.7, the combined $H \to \gamma\gamma$ and $H \to$ ZZ channels cover the mass range above ≈ 115 GeV. Other final states, like $H \to$ WW, suffer from missing energy in the final state and a direct determination of the Higgs mass is not easily possible. The relative mass resolution in the $\gamma\gamma$ final state is about 1% [5, 6] for both ATLAS and CMS. Applying a likelihood method, the Higgs mass can be determined in this channel with a statistical precision below 0.2% using 30 fb^{-1} of data [6].

To estimate the precision on the $M_{\rm H}$ determination in the ZZ final state, a simple Gaussian shape of the peak is assumed and the relative statistical uncertainties are calculated as $\sigma_{M_{\rm H}} = \sigma_{\rm gauss}/\sqrt{N_S}$ and $\sigma_{\Gamma_H} = \sigma_{\rm gauss}/\sqrt{2N_S}$, with the number of Higgs signal events, N_S, in the mass peak interval [6]. This results in uncertainties below 1% on $M_{\rm H}$ for Higgs boson masses up to 500 GeV, and 5–50% on Γ_H, again for 30 fb^{-1}. Such a measurement will put the Standard Model to a very important test. A result which is not in the favourable low-mass range would imply a break-down of the theory at higher energy scales (see Chap. 1).

In an ATLAS study [10] also a possible scenario for the measurement of the Higgs couplings to fermions and bosons has been evaluated. According to Eq. (1.35), they are given by

$$g_W = \frac{M_{\rm W}}{2v}\,, \quad g_Z = \frac{M_{\rm Z}}{2v}\,; \quad g_d = \frac{\sqrt{2}m_f}{v}\,; \quad g_u = -\frac{\sqrt{2}m_f}{v}\,, \tag{7.1}$$

for the W and Z bosons, and up- and down-type fermions. Starting from the measured product of cross-section and branching fraction in the different channels, shown in Fig. 7.8, one can derive eventually also absolute values of the couplings.

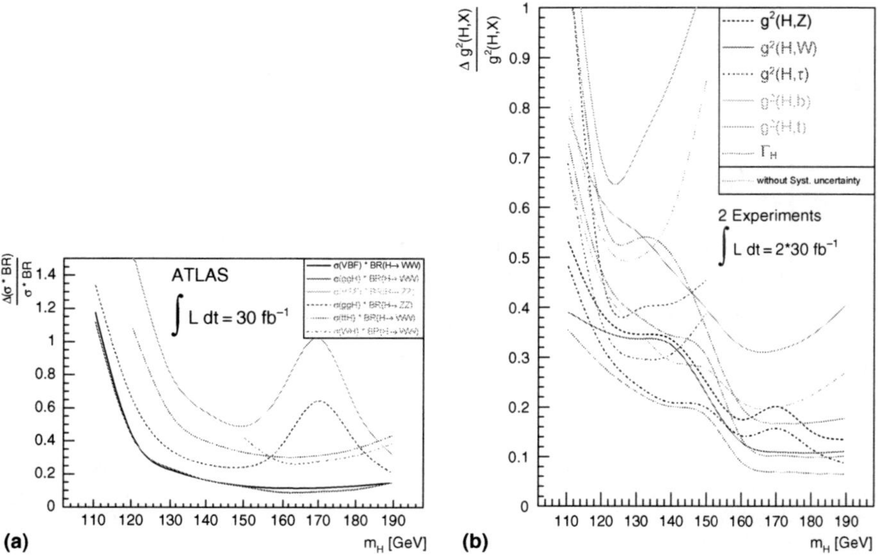

Fig. 7.8 (**a**) Expected precision, including systematic uncertainties, on the measurement of the product $\sigma \times BR$ for the different Higgs decay channels. (**b**) From these, the absolute values of the couplings of the Higgs to fermions and bosons can be derived [10]

This requires however a series of additional assumptions. Under the condition that there is only one CP-even and spin-0 Higgs boson, only Standard Model couplings, and visible branching ratios which follow the Standard Model predictions, the absolute measurement is possible. The result obtained in this restricted scenario is shown in Fig. 7.8. Relaxing the last requirement on the branching fractions, only ratios of Higgs boson couplings are accessible.

7.2.2 Measurement of Spin and CP Quantum Numbers

Most interesting will be the study of the intrinsic properties of the Higgs boson, like its spin and the CP properties. The Standard Model Higgs boson is a scalar particle with positive parity and charge conjugation eigenvalues, P and C, shortly noted as $J^{PC} = 0^{++}$. These parameters can be measured in Higgs production [11] and decay [12] separately. For a quantitative estimate of the sensitivity, the Standard Model Lagrangian is usually extended to include anomalous contributions. They can be from CP-even and CP-odd couplings, CP-violating mixtures of these [13], or from couplings with a Higgs spin structure different from $J = 0$. CP violation in the Higgs sector would be a striking sign for physics beyond the Standard Model. Indeed, the strength of CP violation of the SM, observed only in the $K^0 - \bar{K}^0$ and $B^0 - \bar{B}^0$ systems to date and described by the CKM mixing matrix [14], is not sufficient to explain the baryon/anti-baryon asymmetry in the universe [15]. An additional source of CP violation beyond that of the SM may be needed. An extended Higgs sector together with CP-violating super-symmetry (SUSY) is one possible option of physics beyond the SM that may explain the baryon asymmetry [16]. The measurement of the properties of the Higgs sector and any possible CP violation in Higgs production or decay will be an important question which can be answered once a Higgs boson is discovered [17].

To include different CP structures, the Standard Model HVV vertex function which couples the Higgs field to the vector boson fields, $T^{\mu\nu} = (2M_V/v)g^{\mu\nu}$, is generalised to the tensor [18, 12]:

$$\begin{aligned} T^{\mu\nu} &= a_1(q_1, q_2)g^{\mu\nu} + a_2(q_1, q_2)\left[q_1 \cdot q_2 g^{\mu\nu} - q_2^{\mu} q_2^{\nu}\right] \\ &+ a_3(q_1, q_2)\varepsilon^{\mu\nu\rho\sigma} q_{1\rho} q_{2\sigma} \end{aligned} \tag{7.2}$$

The Lorentz-invariant form factors, a_i, depend on the four-momenta of the weak bosons, q_1 and q_2. The factor a_1 of the first CP-even term appears already in the SM at tree level. The CP-even coupling a_2 and the CP-odd coupling a_3 are from higher dimension operators [19, 20, 18]. They appear first at dimension-5 level and the effective Lagrangian may be written as [18]:

$$\begin{aligned} \mathcal{L}_5 = & \frac{g_{5e}^{HWW}}{\Lambda_{5e}} H W_{\mu\nu}^{+} W^{-\mu\nu} + \frac{g_{5o}^{HWW}}{2\Lambda_{5o}} H \varepsilon_{\mu\nu\rho\sigma} W^{+\rho\sigma} W^{-\mu\nu} \\ & + \frac{g_{5e}^{HZZ}}{2\Lambda_{5e}} H Z_{\mu\nu} Z^{\mu\nu} + \frac{g_{5o}^{HZZ}}{4\Lambda_{5o}} H \varepsilon_{\mu\nu\rho\sigma} Z^{\rho\sigma} Z^{\mu\nu} , \end{aligned} \tag{7.3}$$

where the $\Lambda_{5e/o}$ define the energy scale of the effective theory. The anomalous couplings are then given by

$$a_2(q_1, q_2) = -\frac{2}{\Lambda_{5e}} g_{5e}^{HWW} \; ; \; a_3(q_1, q_2) = -\frac{2}{\Lambda_{5o}} g_{5o}^{HWW} \tag{7.4}$$

and

$$a_2(q_1, q_2) = -\frac{2}{\Lambda_{5e}} g_{5e}^{HWW} \; ; \; a_3(q_1, q_2) = -\frac{2}{\Lambda_{5o}} g_{5o}^{HWW} \tag{7.5}$$

for the *HWW* and *HZZ* vertex. If one assumes the relative strength of W and Z contributions to behave like in the Standard Model, one can additionally require $g_{5e/o}^{HZZ} = \cos^2\theta_{\mathrm{w}} g_{5e/o}^{HWW}$. Setting $a_1 = 0$ and assuming a Higgs mass of 120 GeV, Higgs production rates similar to those in the SM are obtained with $\Lambda_{5e/o} \approx 480$ GeV if only CP-even or only CP-odd couplings are active $\left(g_{5e/o}^{HWW} = 1,\, g_{5o/e}^{HWW} = 0\right)$.

Constraints on these couplings were already set at LEP by the L3 experiment [21]. In the LEP analysis a different effective theory based on dimension-6 operators was used [22–24], and CP-odd terms were neglected:

$$
\begin{aligned}
\mathcal{L}_{\text{eff}} = & \frac{g_2}{2M_{\text{W}}}\left(d\sin^2\theta_{\text{w}} + d_B\cos^2\theta_{\text{w}}\right)\text{HA}_{\mu\nu}\text{A}^{\mu\nu} \\
& + \frac{g_2}{M_{\text{W}}}\left(\Delta g_1^{\text{Z}}\sin 2\theta_{\text{w}} - \Delta\kappa_\gamma\cot\theta_{\text{w}}\right)\text{A}_{\mu\nu}\text{Z}^\mu\partial^\nu\text{H} \\
& + \frac{g_2}{2M_{\text{W}}}\sin 2\theta_{\text{w}}\ (d - d_B)\ \text{HA}_{\mu\nu}\text{Z}^{\mu\nu} \\
& + \frac{g_2}{M_{\text{W}}}\left(\Delta g_1^{\text{Z}}\cos 2\theta_{\text{w}} + \Delta\kappa_\gamma\tan^2\theta_{\text{w}}\right)\text{Z}_{\mu\nu}\text{Z}^\mu\partial^\nu\text{H} \\
& + \frac{g_2}{2M_{\text{W}}}\left(d\cos^2\theta_{\text{w}} + d_B\sin^2\theta_{\text{w}}\right)\text{HZ}_{\mu\nu}\text{Z}^{\mu\nu} \\
& + \frac{g_2\,M_{\text{W}}}{2\,\cos^2\theta_{\text{w}}}\delta_{\text{Z}}\,\text{HZ}_\mu\text{Z}^\mu \\
& + \frac{g_2\,M_{\text{W}}}{M_{\text{Z}}^2}\Delta g_1^{\text{Z}}\left(\text{W}_{\mu\nu}^{+}\text{W}_{-}^{\mu}\partial^\nu\text{H} + h.c.\right) \\
& + \frac{g_2}{M_{\text{W}}}\frac{d}{\cos 2\theta_{\text{w}}}\text{HW}_{\mu\nu}^{+}\text{W}_{-}^{\mu\nu},
\end{aligned}
\tag{7.6}
$$

where g_2 is the $SU(2)_L$ coupling constant. One can identify both sets of CP-even couplings using the relations:

$$
g_{5e}^{HZZ} = \Lambda_{5e}\frac{g_2}{M_{\text{W}}}\left(d\cos^2\theta_{\text{w}} + d_B\sin^2\theta_{\text{w}} + \Delta g_1^{\text{Z}}\cos 2\theta_{\text{w}} + \Delta\kappa_\gamma\tan^2\theta_{\text{w}}\right) \tag{7.7}
$$

$$
g_{5e}^{HWW} = \Lambda_{5e}\frac{g_2}{M_{\text{W}}}\left(d + \frac{M_{\text{W}}^2}{M_{\text{Z}}^2}\Delta g_1^{\text{Z}}\right). \tag{7.8}
$$

The anomalous couplings Δg_1^{Z} and $\Delta\kappa_\gamma$ have already been studied in the analysis of W-pair production at LEP since they also describe possible deviations of the triple-gauge-boson couplings of W bosons with photons and Z bosons [23]. Their measurement in $\text{e}^+\text{e}^- \to \text{WW}$ events was presented in Chap. 3 and they were found to be in very good agreement with the SM expectations. In the general HVV vertex, the additional couplings d and d_B [22] appear. Here, their effects on the Higgs sector shall be further discussed.

The existence of $\text{H}\gamma\gamma$ and $\text{HZ}\gamma$ couplings would actually lead to large $\text{H} \to \gamma\gamma$ and $\text{H} \to \text{Z}\gamma$ branching fractions, which, at tree level, are zero in the Standard Model. These decay modes have complementary sensitivities to the different couplings. In addition, the decay $\text{H} \to \text{WW}^{(*)}$ would be enhanced in the presence of anomalous HWW couplings. This was studied by L3 [21] by interpreting the Higgs search results in the $\text{e}^+\text{e}^- \to HZ$ channels in $H \to \text{f}\bar{\text{f}}$ and $H \to \gamma\gamma$ final states and by extending the signatures to $\text{e}^+\text{e}^- \to H\gamma$ with $H \to \gamma\gamma$, $H \to Z\gamma$, and $H \to \text{WW}^{(*)}$, as well as to the boson-fusion channel $\text{e}^+\text{e}^- \to \text{e}^+\text{e}^- H \to \text{e}^+\text{e}^-\gamma\gamma$. The results are shown in Fig. 7.9. For Higgs masses up to about 100 GeV any anomalous contributions to the HVV vertex are excluded, also for the Δg_1^{Z} and $\Delta\kappa_\gamma$ parameters not shown in the plot. For masses between 100 and 170 GeV, d and d_B couplings larger than 0.5 are found to be incompatible with data at 95% C.L. The

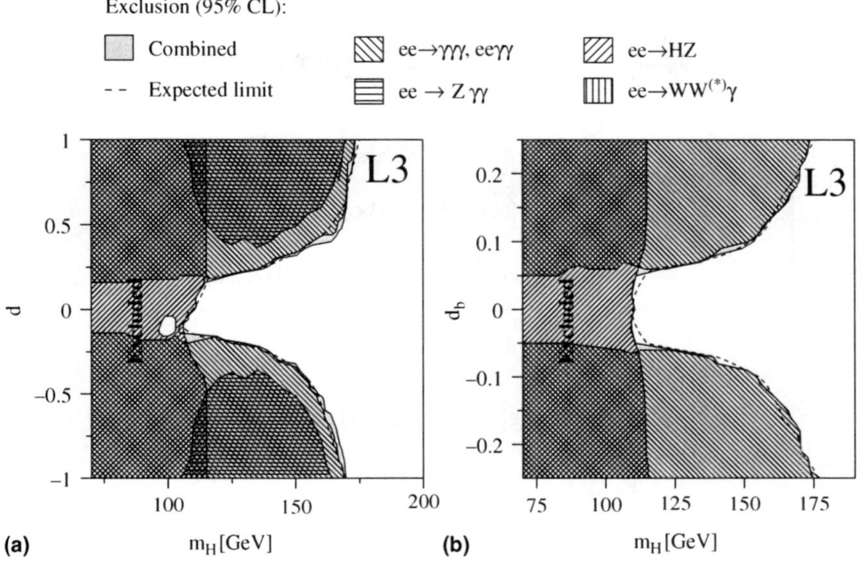

Fig. 7.9 Regions excluded by L3 [21] at 95% C.L. as a function of the Higgs mass for the anomalous couplings: (**a**) d, (**b**) d_B, always setting the other respective couplings to zero

(M_H dependent) limits on Δg_1^Z and $\Delta\kappa_\gamma$ are less stringent than those from the TGC measurements, which are an order of magnitude more sensitive.

Eventually, one can convert the L3 measurements into constraints on the HVV couplings, which yields [11]:

$$g_{5e}^{HWW} \in [-0.78, 0.73]\ ;\quad g_{5e}^{HZZ} \in [-0.63, 0.55] \quad \text{for } M_\mathrm{H} = 120\ \text{GeV}$$
$$g_{5e}^{HZZ} \in [-2.0, 1.5]\ ;\quad g_{5e}^{HZZ} \in [-1.6, 1.3] \quad \text{for } M_\mathrm{H} = 160\ \text{GeV}$$

using Eq. (7.7), assuming classic error propagation and neglecting correlations.

ATLAS studied the measurement of $g_{5e/o}^{HZZ}$ in the VBF channel $\mathrm{qq}H \to \mathrm{qq}\tau^+\tau^-$ with fully leptonic and lepton-hadron decay mode at $M_\mathrm{H} = 120$ GeV as well as in the VBF channel $\mathrm{qq}H \to \mathrm{qqWW} \to \mathrm{qq}\ell\nu\ell\nu$ at $M_\mathrm{H} = 160$ GeV. The distribution of the reconstructed angle between the two forward-tag jets, $\Delta\phi_{\mathrm{jj}}$, was used to evaluate the sensitivity to the anomalous couplings. Figure 7.10 shows the distributions in the three channels for the Standard Model (SM, $a_1 = 1, a_2 = 0, a_3 = 0$), for a scenario with only anomalous CP-even coupling turned on (CPE, $a_1 = 0, a_2 = 1, a_3 = 0$), and finally with only an anomalous CP-odd coupling (CPO, $a_1 = 0, a_2 = 0, a_3 = 1$). It could be shown that from a χ^2 analysis of the $\Delta\phi_{\mathrm{jj}}$ distributions, the CPE and CPO scenarios can be excluded at 95% C.L. with 10 fb^{-1} of data, using only the $H \to \mathrm{WW}$ channel. The low-mass $H \to \tau^+\tau^-$ channel only becomes sensitive beyond 30 fb^{-1}.

The Higgs coupling structure is also easily accessible in the $H \to \mathrm{ZZ} \to \ell^+\ell^-\ell'^+\ell'^-$ decay channel, where the lepton pairs, $\ell^+\ell^-$ and $\ell'^+\ell'^-$, are muon or

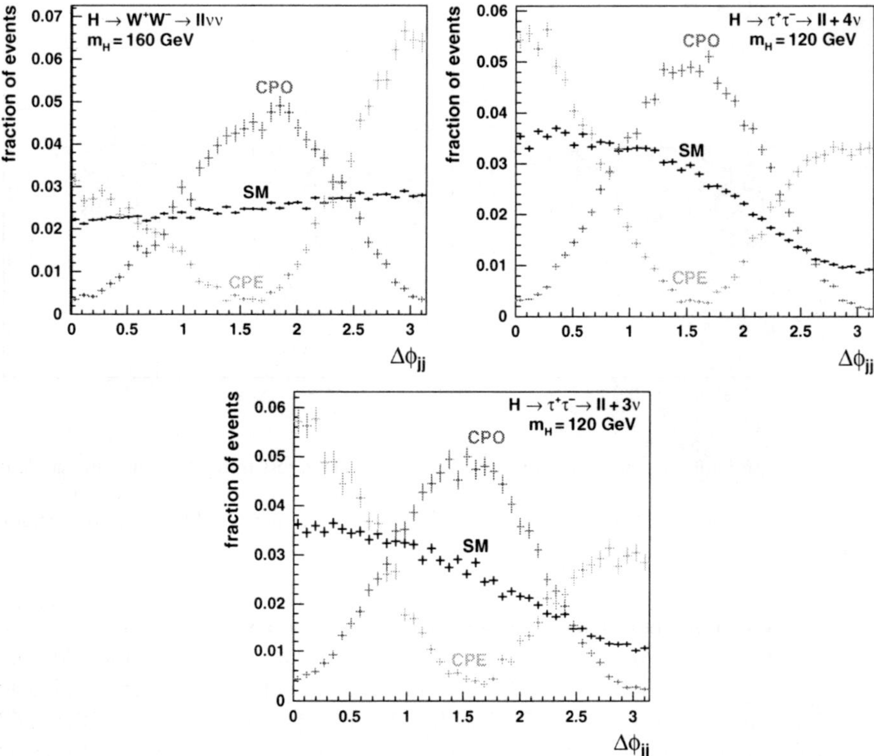

Fig. 7.10 Angle between the forward-tag jets in a high statistics sample of simulated VBF Higgs events for different HVV couplings: the Standard Model (SM) case, and pure anomalous CP-even (CPE) and CP-odd (CPO) couplings [11]

electron pairs. From distributions of the Z decay plane correlations and the lepton angles in the Z rest frames, the helicity structure of the HZZ vertex can be determined. The observables that are most sensitive to the CP properties of the Higgs are

- the angle between the oriented Z decay decay planes, ϕ, in the Higgs rest frame, and
- the cosine of the polar angle of the fermion, $\cos\theta^*$, in the Z rest frame.

Their theoretical distribution expected in the Standard Model and for an anomalous CP-even and CP-odd HZZ coupling are shown in Fig. 7.11.

The ZZ final state was studied by ATLAS [26, 25] and CMS [6, 27]. The theoretical frameworks used in these analyses were based on a coupling structure:

$$T^{\mu\nu}_{HZZ} = \frac{igM_Z}{\cos\theta_W}\left\{Ag^{\mu\nu} + \frac{B}{M_Z^2}p^{\mu}p^{\nu} + \frac{C}{M_Z^2}\varepsilon^{\mu\nu\rho\sigma}p_{\rho}k_{\sigma}\right\}, \qquad (7.9)$$

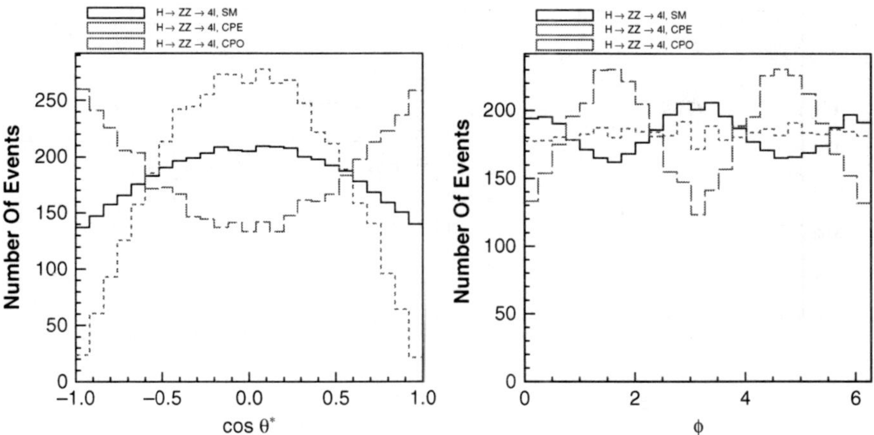

Fig. 7.11 Distribution of the fermion polar angle, θ^*, in the Z rest frame (*left*) and the angle ϕ between the decay planes of the Z bosons (*right*) for the Standard Model (SM), and an anomalous CP-even (CPE) and CP-odd (CPO) vertex. The distributions are shown at Monte Carlo generator level [25]

with the four-momenta of the Z bosons, q_1 and q_2, their sum, $p = q_1 + q_2$, and their difference, $k = q_1 - q_2$. The parameters A, B, and C describe the SM couplings strength and anomalous CP-even and CP-odd contributions, very similar to Eq. (7.2). An alternative model [28] with a different parameterisation of the HZZ vertex is also used in the analysis. This model is integrated in the Pythia [29] event generator. Purely CP-even, CP-odd, and mixed scenarios are available. In the mixed scenario, a parameter η is introduced, and the CP-odd term is varied proportional to η^2, while the interference term scales with η. The η parameter is not directly equivalent to C and contains a small admixture of the CP-even B-term. The limit $\eta \to 0$ corresponds to the Standard Model and $|\eta| \to \infty$ to a pure CP-odd coupling. A mapping $\xi = \text{atan}(\eta)$ projects the full coupling range to the interval $[-\pi, \pi]$.

Usually large data statistics, in the order of 50–100 fb^{-1}, is necessary to measure the anomalous couplings in $H \to$ ZZ decays. CMS simulated signal and background events, which are practically only from ZZ continuum, for a luminosity of 60 fb^{-1}. The likelihood analysis of the shape of the angular distributions and the reconstructed Higgs mass spectrum yields a sensitivity in the order of $\Delta\xi = 0.2$ for $M_{\text{H}} = 200$ GeV, shown in Fig. 7.12, improving with mass and signal statistics to $\Delta\xi = 0.14$ for $M_{\text{H}} = 300$ GeV. This result is in good agreement with expectations for the ATLAS experiment [25], where it could also be shown that systematic effects from resolution and modelling of signal and background can eventually be controlled.

In addition to an anomalous CP structure, the effect of a Higgs-like particle with different spin $J \neq 0$ was evaluated. Clearly, this is mainly to reject this possibility once the Higgs is discovered in the $H \to$ ZZ channel. The main sensitivity to models with $J > 0$ and anomalous HZZ couplings at high Higgs mass is obtained from the analysis of the fermion decay angle, $\cos\theta^*$. This sensitivity can be enhanced by

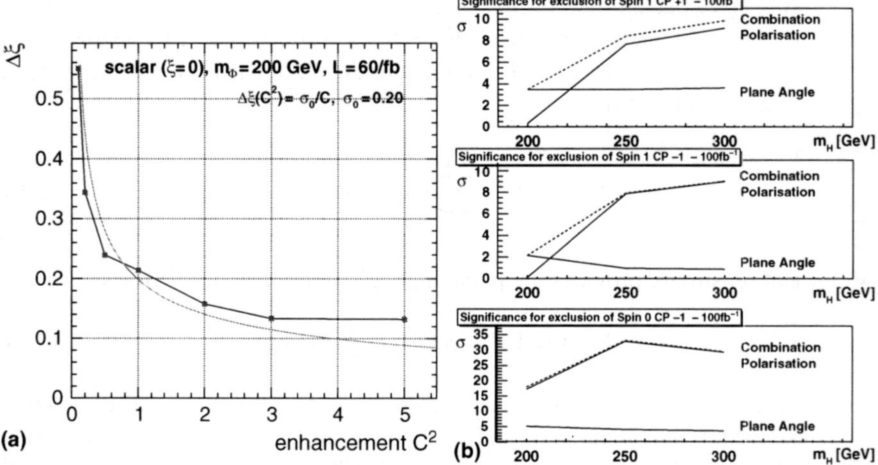

Fig. 7.12 (**a**) Estimated sensitivity to the anomalous HZZ coupling $\xi = \text{atan}(\eta)$ of the Pythia model as a function of a possible signal enhancement factor C for a Higgs mass of 200 GeV and an integrated luminosity of 60 fb^{-1} [27]. The enhancement factor assumes that the Higgs production cross-section scales proportional to C, where for $C = 1$ the SM expectation is obtained. (**b**) Expected significance for excluding spin 1 and anomalous CP-even and CP-odd scenarios for 100 fb^{-1} of ATLAS data [26]

measuring the Z polarisation, assuming a distribution of the decay angle according to [30]:

$$f(\theta) = T(1 + \cos^2\theta^*) + L\sin^2\theta^* \ , \tag{7.10}$$

where T and L are the fraction of transverse and longitudinal polarisation of the Z bosons. A similar method was successfully applied by the KTeV Collaboration who measured the parity of the neutral pion in the four-electron decay, $\pi^0 \to \gamma^*\gamma^* \to e^+e^-e^+e^-$ [31]. The variable

$$R = \frac{L - T}{L + T} \tag{7.11}$$

quantifies the relative difference of longitudinal and transverse polarisation fractions, L and T. For a pure CP-even HZZ coupling a value of $R = +1$ is expected, corresponding to a 100% longitudinal Z polarisation, while $R = -1$ for a pure CP-odd coupling, corresponding to a 100% transverse Z polarisation. In the Standard Model, R is around 0.5, varying with the Higgs mass. Since the background is flat in $\cos\theta^*$ the normalised probability density as a function of θ^* is given by:

$$P(\theta^*) = \frac{\frac{3}{4}N_s\left(\frac{1-R}{3-R}(1 + \cos^2\theta^*) + \frac{1+R}{3-R}\sin^2\theta^*\right) + \frac{1}{2}N_b}{N_s + N_b} \ . \tag{7.12}$$

If the number of background events, N_b, is assumed to be known, the number of signal events is obtained from data as $N_s = N_{\text{data}} - N_b$. An effective value of R is determined from the reconstructed fermion angles, θ_i^*, of each event i, maximising the log-likelihood function

$$\log L = \sum_{i=1}^{N_{\text{data}}} \log P\left(\theta_i^*\right) . \tag{7.13}$$

Additional sensitivity provides the polar angle distribution which follows

$$F(\phi) = 1 + \alpha \cos(\phi) + \beta \cos(2\phi) \tag{7.14}$$

for the HZZ signal (see Fig. 7.11). A likelihood analysis, including also the background shape, of both polar angle and decay plane angle was performed by ATLAS [26], and the result is shown in Fig. 7.12. A CP-odd scalar boson can be well excluded with 100 fb^{-1} of data, while an exclusion of a vector-like particle with spin $J = 1$ is only possible for Higgs masses above 200 GeV. This is again in good agreement with [25].

In the very low-mass region, where $H \to \gamma\gamma$ may be the first discovery channel, only the spin can be constrained using the well-known Landau–Yang theorem [32]. In case this decay mode is observed at the LHC, the Higgs boson must necessarily be of spin-0 nature, or a multiple of $J = 2$.

The measurement of the CP structure in the $H \to \tau^+\tau^-$ decay, which may well be a preferred decay scenario of a low-mass super-symmetric Higgs boson, is possible but challenging. In principle, the polarisation of the τ leptons, measured by analysing the τ decay kinematics, may be used to infer on the Higgs CP and spin structure. But in contrary to the $H \to ZZ \to 4\ell$ channel, the centre-of-mass frame can not be directly reconstructed, which drastically reduces the sensitivity of the decay angular distributions [33]. The Higgs spin and CP structure will therefore be mainly determined in the VBF channels $H \to \tau^+\tau^-$ and $H \to \text{WW}^{(*)}$, as well as in $H \to$ ZZ decays.

In summary, one can conclude that the Standard Model Higgs parameters, like mass, width, branching fractions, as well as spin and CP structure can be probed at the LHC with data samples that correspond to integrated luminosities in the order of 30 fb^{-1}. For further interesting analyses, like the measurement of the Higgs trilinear self-coupling in $H \to HH \to$ WWWW, WWZZ, ZZZZ decays, two orders of magnitude more data is needed [34], which may be delivered by an upgraded super-LHC [35] with instantaneous luminosities of 10^{35} $\text{cm}^{-2}\text{s}^{-1}$.

References

1. For a review, see for example: A. Djouadi, *The Anatomy of Electro-Weak Symmetry Breaking. I: The Higgs Boson in the Standard Model*, Phys. Rept. **457** (2008) 1, hep-ph/0503172; A. Djouadi, *The Anatomy of Electro-Weak Symmetry Breaking. II: The Higgs Bosons in the Minimal Supersymmetric Model*, Phys. Rept. **459** (2008) 1, hep-ph/0503173.

2. ALEPH, DELPHI, L3 and OPAL Collaborations, Phys. Lett. **B 565** (2003) 61.
3. The TEVNPH Working Group, *Combined CDF and DØ Upper Limits on Standard Model Higgs-Boson Production with up to* 4.2 fb^{-1} *of Data*, FERMILAB-PUB-09-060-E, arXiv:0903.4001v1.
4. The ALEPH, DELPHI, L3, OPAL, SLC Collaborations, the LEP Electroweak Working Group, the Tevatron Electroweak Working Group, and the SLD electroweak and heavy flavour groups, *Precision Electroweak Measurements and Constraints on the Standard Model*, CERN-PH-EP/2008-020; arXiv:0811.4682. Updates can be found at http://lepewwg.web.cern.ch/LEPEWWG/.
5. The ATLAS Collaboration, G. Aad et al., *Expected Performance of the ATLAS Experiment Detector, Trigger, Physics*, CERN-OPEN-2008-020, arXiv:0901.0512.
6. The CMS Collaboration, G. L. Bayatian et al., *CMS Physics Technical Design Report, Vol. II: Physics Performance*, CERN/LHCC 2006-021.
7. D. Trocino, *Search for the SM Higgs Boson in the* $H \to ZZ$ *in 4 Leptons at CMS*, CMS Conference Report, CMS CR-2008/108.
8. J. M. Butterworth, A. R. Davison, M. Rubin and G. P. Salam, Phys. Rev. Lett. **100** (2008) 242001; J. M. Butterworth, A. R. Davison, M. Rubin and G. P. Salam, *Jet Substructure as a New Higgs Search Channel at the LHC*, arXiv:0810.0409.
9. Y. L. Dokshitzer, G. D. Leder, S. Moretti and B. R. Webber, JHEP **08** (1997) 001, hep-ph/9707323; M. Wobisch and T. Wengler, hep-ph/9907280.
10. M. Dührssen, S. Heinemeyer, H. Logan, D. Rainwater, G. Weiglein and D. Zeppenfeld, Phys. Rev. **D 70** (2004) 113009; M. Dührsen, *Prospects for the Measurement of Higgs Boson Coupling Parameters in the Mass Range from* $110-190\ GeV/c^2$, ATLAS note, ATL-PHYS-2003-030.
11. Ch. Ruwiedel, M. Schumacher and N. Wermes, *Prospects for the Measurement of the Structure of the Coupling of a Higgs Boson to Weak Gauge Bosons in Weak Boson Fusion with the ATLAS Detector*, ATLAS Scientific Note, SN-ATLAS-2007-060.
12. S. Y. Choi et al., Phys. Lett. **B 553** (2003) 61.
13. R. M. Godbole, D. J. Miller and M. M. Mühlleitner, *Aspects of CP Violation in the H ZZ Coupling at the LHC*, CERN-PH-TH-2007-116, IISC-CHEP-08-07, LAPTH-1195-07, Aug 2007, hep-ph/0708.0458.
14. M. Kobayashi and T. Maskawa, Prog. Th. Phys. **49** 2 (1973) 652.
15. J. M. Cline, Pramana **55** (2000) 33, hep-ph/0003029; M. Dine and A. Kusenko, Rev. Mod. Phys. **76** (2004) 1, hep-ph/0303065; J. M. Cline, hep-ph/0609145.
16. See for example: S. J. Huber, M. Pospelov and A. Ritz, Phys. Rev. D **75** (2007) 036006, hep-ph/0610003.
17. E. Accomando et al., *Workshop on CP Studies and Non-standard Higgs Physics*, CERN-2006-009, hep-ph/0608079 and references therein; R. M. Godbole, Pramana **67** (2006) 835.
18. T. Figy and D. Zeppenfeld, Phys. Lett. **B 591** (2004) 297.
19. W. Buchmüller and D. Wyler, Nucl. Phys. **B 268** (1986) 621; C. J. C. Burges and H. J. Schnitzer, Nucl. Phys. **B 228** (1983) 424; C. N. Leung, S. T. Love and S. Rao, Z. Phys. **C 31** (1986) 433.
20. T. Figy, V. Hankele, G. Klamke and D. Zeppenfeld, Phys. Rev. **D 74** (2006) 095001, hep-ph/0609075v2.
21. L3 Collaboration, M. Acciarri et al., Phys. Lett. **B 489** (2000) 102.
22. G. J. Gounaris, F. M. Renard and N. D. Vlachos, Nucl. Phys. **B 459** (1996) 51.
23. K. Hagiwara, R. D. Peccei and D. Zeppenfeld, Nucl. Phys. **B 282** (1987) 253.
24. B. Grządkowski and J. Wudka, Phys. Lett. **B 364** (1995) 49.
25. A. Straessner, *Prospects for the Measurement of CP Parameters of the Higgs Boson in* $H \to ZZ \to 4\ell$ *Decays with the ATLAS Detector*, ATLAS Internal Note, ATL-COM-PHYS-2008-065.
26. C. P. Buszello, I. Fleck, P. Marquard and J. J. van der Bij, Eur. Phys. J. **C 32** (2004) 209.

27. M. Bluj, *A Study of Angular Correlations in $\Phi \to ZZ \to 2e\mu$*, CMS NOTE 2006/094; Acta Phys. Pol. **B 38** (2007) 739.
28. K. E. Myklevoll, *Possibility of Measuring the CP of a Higgs Boson at the LHC*, M.Sc. Thesis, University of Bergen, 2002.
29. T. Sjöstrand, P. Edén, C. Friberg, L. Lönnblad, G. Miu, S. Mrenna and E. Norrbin, Comp. Phys. Comm. **135** (2001) 238; For colour reconnection and Bose-Einstein correlation studies PYTHIA version 6.121 is used. T. Sjöstrand, *Recent Progress in PYTHIA*, Preprint, LU-TP-99-42, hep-ph/0001032.
30. C. P. Buszello, *Bestimmung der Kopplungsstruktur des Higgs-Bosons mit dem Atlas-Detektor*, Ph.D. Thesis, Freiburg University, 2004.
31. The KTeV Collaboration, E. Abouzaid et al., Phys. Rev. Lett. **100** (2008) 182001, arXiv:0802.2064.
32. L. D. Landau, Dokl. Akad. Nauk., USSR **60**(1948) 207; C. N. Yang, Phys. Rev. **77** (1950) 242.
33. G. R. Bower, T. Pierzchala, Z. Was and M. Worek, Phys. Lett. **B 543** (2002) 227; Z. Was and M. Worek, Acta Phys. Pol. **B 33** (2002) 1875, hep-ph/0202007v2.
34. A. Blondel, A. Clark and F. Mazzucato, *Studies on the Measurement of the SM Higgs Self-couplings*, ATLAS Physics Note ATL-PHYS-2002-029.
35. G. Darbo et al., *Outline of R&D Activities for ATLAS at an Upgraded LHC*, ATLAS Note ATL-COM-GEN-2005-002.

Chapter 8
Summary and Conclusion

The Standard Model of electroweak interactions is currently in very good agreement with high-precision experimental data, as detailed in Chaps. 3 and 4. Cornerstones are the precise determination of mass and width of the Z and W bosons, which currently yield:

$$\begin{aligned}
M_Z &= 91.1875 \pm 0.0021 \text{ GeV} , \\
\Gamma_Z &= \ \ 2.4952 \pm 0.0023 \text{ GeV} , \\
M_W &= 80.399 \pm 0.023 \text{ GeV} , \\
\Gamma_W &= \ \ 2.098 \pm 0.048 \text{ GeV} ,
\end{aligned}$$

combining LEP and Tevatron results. By studying in detail the Z decays to fermions, the corresponding coupling constants and partial decay widths were measured and the electroweak mixing angle together with the ρ-parameter were extracted as

$$\begin{aligned}
\sin^2 \theta^{\ell}_{\text{eff}} &= 0.23153 \pm 0.00012, \\
\rho_\ell &= 1.0050 \pm 0.0010.
\end{aligned}$$

These results differ significantly from the tree level predictions of the Standard Model, showing clearly the necessity to include radiative corrections. Their analysis indicates that the Standard Model Higgs boson should be lighter than 186 GeV when evaluating a large set of electroweak data.

At LEP energies above the Z pole, the electroweak gauge bosons were produced singly or in pairs which allowed a detailed study of the triple and quadruple gauge boson couplings. In both, the neutral and charged sector, they were found to be in good agreement with the non-abelian $SU(2) \times U(1)$ gauge structure of the Standard Model. In particular, the W boson couplings to the photon and Z boson were determined as:

$$\begin{aligned}
g_1^Z &= \ \ 0.991^{+0.022}_{-0.021} , \\
\kappa_\gamma &= \ \ 0.984^{+0.042}_{-0.047} , \\
\lambda_\gamma &= -0.016^{+0.021}_{-0.023} ,
\end{aligned}$$

in perfect agreement with the theoretical prediction.

A. Straessner, *Electroweak Physics at LEP and LHC*, STMP 235, 209–211,
DOI 10.1007/978-3-642-05169-2_8, © Springer-Verlag Berlin Heidelberg 2010

The measurement of the W mass at LEP and the control of the corresponding systematic uncertainties would not have been possible without the detailed understanding of Final State Interactions in hadronic W-pair decays. The effects of Bose-Einstein correlations between the decay products of two hadronically decaying W bosons were found to be small or absent. However, there is a 51% probability for Colour Reconnection between these decay products. The LEP W-mass analyses were therefore optimised to reduce the sensitivity of the mass measurement to Colour Reconnection effects.

The large collection of electroweak data, including top mass and low-Q^2 measurements, provides constraints to the last missing piece in the Standard Model, the mass of the Higgs boson, though not proving its existence. This will only be achieved by direct searches which are currently ongoing at the Tevatron and which will be intensified once the LHC will start collecting data.

The LHC collider and the experiments ATLAS and CMS are now completed and ready to take data. Subdetectors for measuring electrons, muons, taus and jets with good precision were built. They will be able to trigger on and measure the different particles up to the design luminosity of 10^{34} cm^{-2}s^{-1} at centre-of-mass energies between 7 and 14 TeV. The road-map to achieve the necessary detector performance is laid out, and analysis frameworks to measure the detector parameters from data are being prepared.

ATLAS and CMS are expected to improve the precision measurements of the masses of the W boson and of the top quark, with uncertainties below 10 MeV and 1 GeV respectively, once the detectors are understood and their performances optimised. By studying Z boson decays to electron-positron pairs the electroweak mixing angle, $\sin^2\theta^{\ell}_{\text{eff}}$, can be determined with an accuracy of about 2×10^{-4}, close to the LEP and SLC results. The sensitivity to triple gauge boson couplings is enhanced at the LHC due to the high centre-of-mass energy that will be reached. Anomalous contributions to the TGCs will be tested at the per cent level. Constraints from electroweak data for theoretical models will therefore be narrowed further.

Once the LHC running starts and enough data is collected, the ATLAS and CMS experiments will eventually clarify if the Standard Model Higgs boson exists. In the complete theoretically and experimentally possible mass range of $114.4 \text{ GeV} < M_{\text{H}} < 1 \text{ TeV}$ the combined analyses of ATLAS and CMS will be able to discover the Higgs boson with about $2 - 20$ fb^{-1} of data depending on the Higgs mass. With some more luminosity collected, the properties of the Higgs boson, like mass, width, couplings, spin and CP structure can be measured.

However, there are arguments, of which some are mentioned in Chaps. 1 and 4, that the theory may need to be extended beyond the well-working Standard Model. Electroweak symmetry breaking, which necessarily appears at the TeV scale, may be induced by super-symmetry, composite Higgs models, or theories with strongly interacting vector-bosons. These signatures may appear even before a Higgs boson is seen, for example as heavy Z' or W' bosons or light super-symmetric particles.

At the advent of the LHC start, the LHC experiments are well prepared to further test the Standard Model and to search for new physics beyond it. Very rare processes and detailed studies of the findings will be possible at the upgraded LHC, the sLHC, with a factor of 10 increase in instantaneous luminosity. The ATLAS and CMS detector communities are already now investigating further improvements of their tracking and forward detectors, and their trigger and detector readout systems to be prepared for an even more challenging background environment.

Complementary measurements to the LHC in the Higgs, electroweak, and super-symmetric sector will be possible with a future International Linear Collider [1–3]. In e^+e^- collisions up to 500 TeV, the Standard Model Higgs and top quark masses can be measured with about 50 MeV precision. The HVV couplings can be derived with a few percent uncertainty from the Higgs-strahlung and boson-fusion production cross-sections. The top Yukawa coupling to the Higgs boson is measurable in $t\bar{t}H$ production, and the ratios of the other $f\bar{f}H$ couplings are accessible through the different branching fractions $Br(H \rightarrow f\bar{f})$. If spin and CP parameters of the Higgs boson will not have been determined by the LHC, the ILC experiments will be able to measure those, e.g., by scanning the threshold region of $e^+e^- \rightarrow ZH$ production and by studying the ZH production and decay kinematics. The ILC will also allow a more detailed measurement of the properties of possible super-symmetric particles that may be discovered at the LHC. The tandem of ILC precision measurements and LHC discovery potential in Higgs and SUSY physics can therefore fulfil a similar task as the $Sp\bar{p}S$ and LEP experiments in the past in case of W and Z discovery and precision physics.

In the very near future it is however guaranteed that the LHC experiments, once having analysed the first few 100 pb^{-1} to fb^{-1} of data and especially when running at LHC design luminosity and beyond, will certainly change the landscape in particle physics. They will provide interesting insights into physics at the TeV energy scale and will eventually shed light on the mechanism of electroweak symmetry breaking.

References

1. A. Djouadi, J. Lykken, K. Mönig, Y. Okada, M. Oreglia, S. Yamashita et al., *ILC Reference Design Report Volume 2 – Physics at the ILC*, arXiv:0709.1893.
2. N. Phinney, N. Toge, N. Walker et al., *ILC Reference Design Report Volume 3 – Accelerator*, arXiv:0712.2361.
3. T. Behnke, C. Damerell, J. Jaros, A. Myamoto et al., *ILC Reference Design Report Volume 4 – Detectors*, arXiv:0712.2356.

Index

© Springer-Verlag Berlin Heidelberg 2010

Index

A. Straessner, *Electroweak Physics at LEP and LHC*, STMP 235, 213–214,
DOI 10.1007/978-3-642-05169-2, © Springer-Verlag Berlin Heidelberg 2010

Lightning Source UK Ltd.
Milton Keynes UK
03 April 2010

152243UK00001B/64/P

9 783642 051685